Contents

How To Use This Book

The *Year 3 Targeting HASS Activity Book* is filled with exciting stimulus materials to assist you in developing a greater understanding of the world you live in—its past and its present. The activities have been designed to encourage you to share opinions, think creatively, analyse ideas and information and respond in a variety of ways.

Each unit delivers content from the knowledge and understanding strand of the Australian Curriculum and supports the key inquiry questions in each subject area. The units also allow you to develop inquiry skills through:

Researching

Students identify and collect information, evidence and data from primary and secondary sources.

Questioning

Students develop questions about events, people, places, ideas, developments and issues to guide their investigations and satisfy their curiosity.

Analysing

Students explore information, evidence and data to identify and interpret features, patterns, trends and relationships, key points, facts and opinions and points of view.

Communicating

Students present ideas, findings, viewpoints, judgements and conclusions in digital and non-digital forms for different audiences and purposes.

Evaluating and Reflecting

Students propose explanations for events, developments and issues, draw evidence-based conclusions and use criteria to make informed decisions and judgements.

This book is organised into the Year 3 curriculum areas, as follows:

- **Units 1–14: HISTORY**
- **Units 15–26: GEOGRAPHY**
- **Units 27–32: CIVICS AND CITIZENSHIP**

TARGETING HASS 3 © PASCAL PRESS ISBN: 9781925726046

The ***Targeting HASS Activity Book*** series is designed to be a flexible learning resource. The units do not have to be completed in numerical order. Your teacher may direct you to complete units and activities from one learning area before moving on to a different learning focus. In this way, the *Targeting HASS* series will complement any school's scope and sequence.

The seven **Australian Curriculum's General Capabilities** listed below are embedded in the units of this book:

- Literacy
- Numeracy
- Information and Communication Technology (ICT) Capability
- Critical and Creative Thinking
- Personal and Social Capability
- Ethical Understanding
- Intercultural Understanding

A number of the activities and questions are open-ended. This allows for student challenge and differentiation. The questions are written to encourage you to think deeply about topics and provide extended responses. For some open-ended questions, answers are not provided. These questions are best graded through peers marking each other's work, or through student–teacher conferences. There are brief sample responses to guide the assessment of your work for some open-ended questions.

Why are national symbols, such as our coat of arms and flag, important to different groups and individuals?

Evaluating and Reflecting

Do we have too many rules in public places? What is one rule that you think we don't need any more? Explain why.

The final section features eight **assessment** activities:

- 3 History
- 3 Geography
- 2 Civics & Citizenship.

Each assessment activity relates to a key inquiry question. Make sure you have completed the units for that inquiry question before tackling the Assessment.

Additional inquiry skills covered by each unit

		ACARA CODE	WA CODE	VIC CODE
Questioning	Pose questions for investigations	ACHASS1052	WAHASS27	
Researching	Research, locate and collect data	ACHASS1053	WAHASS28	VCHHC067
	Record, sort and represent data	ACHASS1054	WAHASS29	VCGGC074
	Sequence information	ACHASS1055		VCHHC066
Analysing	Examine information and distinguish facts from opinions	ACHASS1056	WAHASS33	VCHHC068
	Interpret data	ACHASS1057	WAHASS32	VCGGC076
Communicating	Present ideas, findings and conclusions	ACHASS1061	WAHASS34 WAHASS37	VCGGC075
Evaluating and Reflecting	Draw simple conclusions based on analysis of information	ACHASS1058	WAHASS35	
	Interact with others to share points of view	ACHASS1059	WAHASS36	
	Reflect on learning and consider possible effects of proposed actions	ACHASS1060	WAHASS39	

Curriculum correlations

ACARA CODE: DESCRIPTION – HISTORY
ACHASSK062 The importance of Country/Place to Aboriginal and/or Torres Strait Islander Peoples who belong to a local area
ACHASSK063 How the community has changed and remained the same over time and the role that people of diverse backgrounds have played in the development and character of the local community
ACHASSK064 Days and weeks celebrated or commemorated in Australia (including Australia Day, ANZAC Day and National Sorry Day) and the importance of symbols and emblems
ACHASSK065 Celebrations and commemorations in places around the world (for example Chinese New Year in countries in the Asian region) including those that are observed in Australia (for example Hanukkah, the Moon festival and Ramadan)
ACARA CODE: DESCRIPTION – GEOGRAPHY
ACHASSK066 The representation of Australia as states and territories and the diverse characteristics of their places
ACHASSK067 The location of Australia's neighbouring countries and the diverse characteristics of their places
ACHASSK068 The main climate types of the world and the similarities and differences between the climates of different places
ACHASSK069 The similarities and differences between places in terms of their types of settlement, demographic characteristics and the lives of people who live there, and people's perceptions of these places
ACARA CODE: DESCRIPTION – CIVICS AND CITIZENSHIP
ACHASSK070 The importance of making decisions democratically
ACHASSK071 Who makes rules, why rules are important, and the consequences of rules not being followed
ACHASSK072 Why people participate with communities and how students can actively participate and contribute
NSW HISTORY STAGE 2 CURRICULUM OUTCOMES
HT2-1 identifies celebrations and commemorations of significance in Australia and the world
HT2-2 describes and explains how significant individuals, groups and events contributed to changes in the local community over time
HT2-5 applies skills of historical inquiry and communication
NSW GEOGRAPHY STAGE 2 CURRICULUM OUTCOMES
GE2-1 examines features and characteristics of places and environments
GE2-2 describes the ways people, places and environments interact
GE2-3 examines differing perceptions about the management of places and environments
GE2-4 acquires and communicates geographical information using geographical tools for inquiry
VICTORIA HISTORY LEVELS 3 & 4 CURRICULUM OUTCOMES
VCHHK072 The significance of Country and Place to Aboriginal and Torres Strait Islander peoples who belong to a local area
VCHHK073 A significant example of change and a significant example of continuity over time in the local community, region or state/ territory
VCHHK075 One significant narrative, myth or celebration from the past
VCHHK076 Significance of days and weeks celebrated or commemorated in Australia and the importance of symbols and emblems, including Australia Day, ANZAC Day, Harmony Week, National Reconciliation Week, NAIDOC week and National Sorry Day
VCHHK077 Significance of celebrations and commemorations in other places around the world
VICTORIA GEOGRAPHY LEVELS 3 & 4 CURRICULUM OUTCOMES
VCGGK069 Reasons why some places are special and some places are important to people and how they can be looked after
VCGGK074 Collect and record relevant geographical data and information from the field and other sources
VCGGK076 Interpret maps and other geographical data and information to develop identifications, descriptions, explanations and conclusions, using geographical terminology including simple grid references, compass direction and distance
VCGGK078 Location of Australia's neighbouring countries and the diverse characteristics of their places
VICTORIA CIVICS & CITIZENSHIP LEVELS 3 & 4 CURRICULUM OUTCOMES
VCCCG002 Identify how and why decisions are made democratically in communities
VCCCG004 Explain how and why people make rules
VCCCG006 Investigate why and how people participate within communities and cultural and social groups
VCCCG007 Describe the different cultural, religious and/or social groups to which they and others in the community may belong

History														Geography												Civics and Citizenship						History Assessment			Geography Assessment			Civics & Citizenship Assessment	
1	2	3	4	5	6	7	8	9	10	11	12	13	14	15	16	17	18	19	20	21	22	23	24	25	26	27	28	29	30	31	32	1	2	3	1	2	3	1	2
✔	✔	✔																																					
			✔	✔	✔	✔	✔	✔																										✔					
									✔	✔																						✔							
											✔	✔	✔																				✔						
														✔	✔	✔	✔																		✔				
																		✔	✔	✔	✔															✔			
																						✔	✔	✔	✔												✔		
																						✔	✔	✔	✔												✔		
																										✔	✔												
																												✔	✔									✔	
																														✔	✔								✔
			✔	✔	✔	✔	✔	✔	✔	✔	✔	✔	✔																										
✔	✔	✔	✔	✔	✔	✔	✔	✔	✔	✔																													
✔	✔			✔	✔					✔			✔																										
														✔	✔	✔	✔	✔	✔	✔	✔	✔	✔	✔	✔														
																						✔	✔	✔	✔														
																						✔	✔	✔	✔														
														✔	✔			✔	✔			✔		✔	✔														
✔	✔	✔																																					
			✔	✔	✔	✔	✔	✔	✔	✔																						✔		✔					
✔	✔	✔																																					
											✔	✔	✔																				✔						
											✔	✔	✔																				✔						
														✔	✔	✔	✔																		✔				
																						✔	✔	✔	✔												✔		
																						✔	✔	✔	✔														
																		✔	✔	✔	✔															✔			
																										✔	✔												
																												✔	✔									✔	
																														✔	✔								✔
																														✔	✔								✔

Home

When Captain James Cook claimed Australia as British territory in 1770, he declared it *terra nullius*, which means it was 'nobody's land'.

Even though they had met with a number of different groups of Aboriginal people during their exploration of the east coast, the British believed that Australia was not settled because there was no evidence of houses, towns, roads or farms.

In Bruce Pascoe's book *Young Dark Emu*, he says, "Britain used this reasoning to claim Australia."

Bruce, an Aboriginal writer from the Bunurong clan in Victoria, says that although many people believed that Aboriginal Australians were nomadic and lived in simple temporary structures, they were incorrect.

He explains this by using extracts from explorers' diaries and journals which are filled with notes about Aboriginal housing all over Australia – even those parts of the country that people think of as being harsh and difficult to live in.

One example of this evidence against *terra nullius* comes from the diaries of the explorer Thomas Mitchell. Mitchell wrote about his surprise when he saw large villages with dozens of houses:

[Some huts] ... being large, circular; and made of straight rods meeting at an upright pole in the centre; the outside had first been covered with bark and grass, and the entirety coated over with clay. The fire appeared to have been made nearly in the centre; and a hole at the top had been left as a chimney.

Source: Mitchell, T.L.(1839)

By counting the number of houses, Mitchell estimated that almost 1000 people lived in the village.

Even though the evidence around the village suggested that people had been living there for a very long time, they had all left by the time Mitchell and his team of explorers arrived. Bruce Pascoe believes that permanent settlements such as this village are clear evidence of Aboriginal people's need and use of a reliable and known food source, possibly as a result of farming and agriculture.

Early settlers called these Aboriginal houses 'humpies'.

Illustration: Sefco Obucina. **Reference:** *Young Dark Emu*, Bruce Pascoe, Magabala books

TARGETING HASS 3 © PASCAL PRESS ISBN: 9781925726046

Researching

Which Aboriginal peoples inhabited the area in which you live? What did they call themselves? Find out some details about the way they lived.

https://aiatsis.gov.au/explore/articles/aiatsis-map-indigenous-australia

Questioning

Use the text to complete the **KWL** chart.

What did I **know** about Aboriginal life in Australia?	
What do I still **want** to know?	
What did I **learn** about Aboriginal life in Australia?	

Analysing

Are these statements **true** or **false**?

1 Australia belonged to no one when the British settlers arrived. ____________

2 There was lots of evidence to show that Aboriginal people did not build any houses or shelters. ____________

3 Thomas Mitchell did not expect to see Aboriginal houses. ____________

4 Aboriginal villages were very new and probably only happened because of British settlement. ____________

5 Aboriginal settlements were signs that the people had a reliable food source. ____________

Communicating

Imagine you are a newspaper journalist reporting the experiences of Thomas Mitchell. What would your headline be? Write a brief newspaper article on the topic of Aboriginal housing.

Evaluating and Reflecting

Why did the writer include the extract from Thomas Mitchell's diary? What effect does it have on the text?

First Peoples: Queensland and South Australia

HISTORY

QUEENSLAND

Many archaeological sites that have been excavated in Queensland indicate Aboriginal activity dating back 15 000 to 30 000 years. Queensland's environment as we know it today was very different back then, with cooler temperatures and much of the state covered by forests and grasslands. Like the Koorie peoples of New South Wales and Victoria, Aboriginal peoples from Queensland who identify regionally refer to themselves as Murri.

In 1770, the HMS *Endeavour* struck the Great Barrier Reef off the coast of Northern Queensland. It was beached for seven weeks while repairs were undertaken. During that time, Cook interacted on several occasions with the Guugu Yimithirr people, who occupied the Hope Vale region on the Cape York Peninsula.

These were the first meetings between Cook and Aboriginal peoples. The Guugu Yimithirr referred to themselves as the 'saltwater people' and their coastal location provided varied food sources. They travelled to nearby islands to gather root vegetables, clamshells and seagull eggs; and to hunt for turtles, fish and dugongs.

SOUTH AUSTRALIA

The traditional owners of the Adelaide Plains, the Kaurna people, have occupied South Australia for at least 40 000 years. Their lands extend towards Crystal Brook in the north, along the coast to Cape Jervis in the south and to the Mount Lofty Ranges in the east.

Understanding their environment was of great importance to the Kaurna people, and they practised sustainable land management for many generations.

They combined their knowledge of seasonal harvesting with fire-stick farming, which involved lighting fires to aid hunting and modify the landscape. Burning old grass and vegetation encouraged new growth and also created new grazing land for animals. The excellent land management skills of the Kaurna people meant that food was plentiful, so there was much time left in the day for storytelling and other cultural activities.

Source: *Blake's Australian History Guide*, p. 19, Pascal Press

Research

Marni ninna budni means 'welcome' in the Kaurna language. Hear it at: *www.kaurnaplacenames.com/index.php*
What is the language of the traditional owners of the land where you live? What is their word for 'welcome'? How do they welcome people to their Country?

____

____

____

____

____

____

____

____

Questioning

Write questions that have these answers.

Answer: Murri

Question:

____

____

____

Answer: Saltwater People

Question:

____

____

____

Answer: Sustainable land management

Question:

____

____

____

Analysing

Complete the table below to show the changes in the Queensland environment over time.

Then (30 000 years ago)	**Now**

Communicating

The Kaurna people had time for storytelling and art. What would you paint or draw to show others your way of life? Discuss this with a partner and share your ideas.

Evaluating and Reflecting

Was fire-stick-farming a clever practice? Why?
What planning and organisation would have been needed?

____

____

____

UNIT 3

HISTORY

Dreaming Story: Why the crocodile rolls

Long ago, in the Dreaming, the people lived together happily. The fathers showed their sons the way to become a man. The mothers prepared the evening meals of fish and mud crabs. The little girls played along the shore. This season, there was plenty of food and everyone was content – except a girl named Min-na-wee.

Ever since she was a little girl, Min-na-wee had been a troublemaker. She never said nice things about the other girls and she liked to start fights. Min-na-wee almost never smiled, and her face was hard and scaly. The elders of the tribe told her mother that if Min-na-wee didn't stop being so mean, bad things would happen.

But as the years went on, nothing changed. One day, the young women were getting ready to become brides. The elders showed the men which woman they would marry. At the end of the ceremony, Min-na-wee was left all alone. Nobody had chosen her. This made her very angry!

Now Min-na-wee made even more trouble than before and this made everyone in the tribe upset. So the elders got together and agreed that Min-na-wee must be punished. Several men from the tribe went up to Min-na-wee and grabbed her, but she started rolling around and around in the dirt and escaped. Min-na-wee ran to the edge of the water and she called to the evil spirits to help her. They changed her into a nasty animal so she could attack her own tribe.

Min-na-wee became a crocodile. She lay in the water until one of the men who had tried to punish her came by. Then she grabbed him and rolled him over and over until she was happy that he had been punished for what he had done to her.

That is why today, when a crocodile catches its prey, it will roll around and around in the water.

NOTE: This Dreaming story is from the Kwini people of Western Australia near Broome. It is retold from a legend told to Frank Martin by his grandfather Jiller-rii, a Kwini elder.

Source: Australian History Centres: *Lower Primary*, p. 15, Blake Education

Research

Dreaming stories tell of the ancestor spirits who created the land and its features. They also tell us how to behave towards each other.

Investigate other Dreaming stories from around Australia. Find three stories for each category in the table.

Purpose: explain how natural features were created	Purpose: explain how people should behave

Questioning

Write three questions you could ask a friend to find out if they understood the story, *Why the crocodile rolls*.

1 ______________________________

____

____

2 ______________________________

____

____

3 ______________________________

____

____

Analysing

What could the tribe have done when Min-na-wee was young to stop her from being so mean?

____

____

____

____

____

____

Communicating

Create a story cube to use with your friends. Put a part of the story on each side of the cube. For example: setting, characters, problem.

Or, put questions such as: What was your favourite part? Why do you think Min-na-wee was so mean?

Use your cube with your friends to share your understanding of the story.

Evaluating and Reflecting

If Min-na-wee knew that she would turn into a crocodile because of her behaviour, do you think she would have acted differently? Would she have changed before it was too late? Why do you think this?

____

____

____

UNIT 4

Looking After Country

Maintaining the balance between people and the natural world is called 'looking after country' by the Gagudju people of Kakadu in the Northern Territory. The people's duty to look after the land takes many forms and is lived every day of their lives.

HISTORY

A Gagudju elder sings the Indjuwanydjuwa song cycle. An image of the creator is painted on the cave wall behind him.

Part of this duty is to teach the sacred knowledge to each new generation. This learning takes place over a lifetime, through ceremonies, dance, the colours and designs of their artwork, and through the teachings of the tribal elders.

One of these ceremonies is the song cycle which honours Indjuwanydjuwa, the creator of the world. The ceremony takes place in a cavern overlooking the rock that is the symbol of this creator being. The rock is the focus of the area's life force. By singing the songs and sharing the stories, the elders have kept the traditions alive for more than 2500 years.

Indjuwanydjuwa taught the Gagudju how, when and where to hunt certain animals. Some of the other ways in which the Gagudju take care of the land and their relationship with it are simply to be in the country, to hunt and gather food in it, and to burn in strict accordance with ancient tradition. The Gagudju believe that fires lit in the right season will cleanse the country.

The seasons, the plants, the animals and the landforms of Kakadu – all linked with the stories of the Gagudju's Dreamtime – are at the heart of the people's connection to the land, their 'Country'. In Kakadu, it is still possible to see, to hear, to smell and to feel Australia as it was more than 200 years ago.

Source: Sharing Country: *Kakadu*, pp. 72–76, Steve Parish

How has our community changed?
What has been lost and what features have been retained?

UNIT 4

Research

Investigate the role of Aboriginal elders in retaining (keeping alive) Aboriginal cultures and traditions. Write five facts you found out.

1 ______________________________

____

2 ______________________________

____

3 ______________________________

____

4 ______________________________

____

Analysing

How important are stories and art in our society? Why are they important? What would our community be like today if we did not share experiences and ideas through stories and art?

____

____

____

____

____

Questioning

After reading the text, what questions would you ask a Gagudju elder? What would you like to know about Gagudju culture, traditions and history? Write three questions.

1 ______________________________

____

____

2 ______________________________

____

____

3 ______________________________

____

____

Communicating

Create an acrostic poem or a piece of collage art to represent your culture and community.

Share it with your friends and discuss how it represents your beliefs and experiences.

Evaluating and Reflecting

A time capsule is a place where you put important items, memories of special experiences and souvenirs from places. Its purpose is to retain knowledge so it won't be lost over time.

Imagine that you are creating a time capsule. List six things that you would put in your time capsule, and explain why.

____

____

____

____

UNIT 5

HISTORY

Sydney Harbour Bridge

The bridge before it was finished

The Sydney Harbour Bridge took eight years to build and was opened on 19 March 1932. It allowed people to cross from one side of Sydney Harbour to the other.

The bridge today

Crowds of people crossing the bridge on opening day

Opening day was a big celebration.

There was a colourful pageant with 27 floats, marching bands, pipe bands, 600 school children, scout groups, soldiers and different types of vehicles all parading across the bridge.

A special song was written for the opening. It was called *The Bridge We've Been Waiting For.*

The Bridge We've Been Waiting For
Words and music by Jack Lumsdaine

See the people in the Street
Hear the tramping of the feet, today's a holiday.
Flags are flying everywhere
Carnival is in the air
Down old Sydney way.
Bands are playing
Here they come, listen to the big bass drum,
Listen to the hip hooray.
When the planes up in the sky, fly above the Span
So high
They're opening the Bridge today.
What is everybody saying?
It's the bridge we've been
Waiting for.
Like a giant of steel, at last the dream is real.
Doesn't it make you feel you love old Sydney more?
Right across the dear old Harbour
There's an ever open door.
Australia's Sons let us rejoice
It's the Bridge we've been
Waiting for...

Source: Australian History Centres: *Lower Primary*, p. 21, Blake Education

TARGETING HASS 3 © PASCAL PRESS ISBN: 9781925726046

Research

Identify a special landmark near where you live. What is it? When was it built? Why was it built? How has it changed over time?

____

____

____

____

____

____

____

____

Questioning

Write three questions you have about what life was like in Sydney when the bridge was being built.

1 ____

____

____

2 ____

____

____

3 ____

____

____

Analysing

Use the text to create a Venn diagram comparing the Sydney Harbour Bridge in 1930 and now.

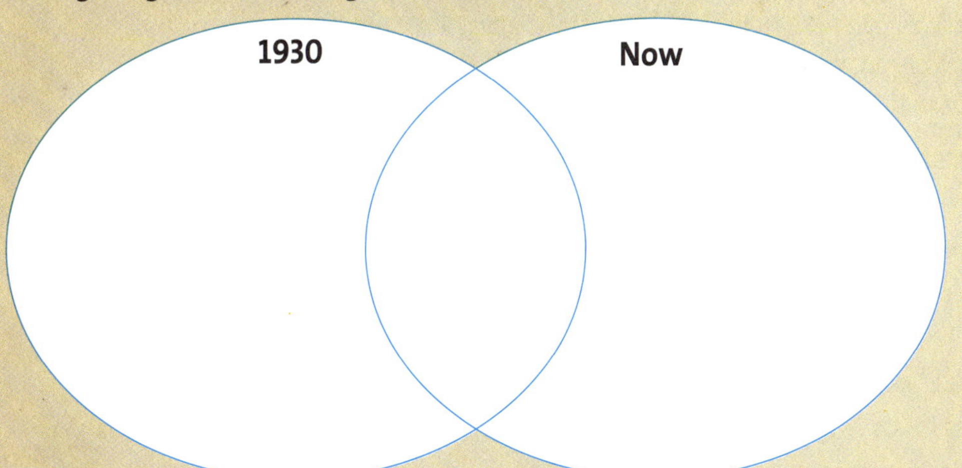

Communicating

Imagine you were at the opening of the Sydney Harbour Bridge in 1932. Write a podcast to report on the event. Record it to share with others.

Evaluating and Reflecting

What makes a constructed landmark, like the Sydney Harbour Bridge, so important?

____

____

The Sydney Harbour Bridge is nicknamed the coat hanger – can you tell why? Why do we give nicknames to famous structures?

____

____

Case study: The Snowy Scheme

Built from 1949 to 1974, The Snowy Mountains Scheme was the largest engineering project ever undertaken in Australia. About 70 per cent of the people who built it were migrants.

The Snowy Scheme was designed to collect water from the Snowy and Eucumbene Rivers into dams and then use it to generate electricity. During the planning stages, it became obvious that there were not enough unemployed people in Australia to work on the project.

The government's Assisted Passage Scheme helped pay for many Europeans to migrate to Australia to work on the Snowy Scheme. At first, most of them came from Britain, but by the early 1950s other Europeans followed.

Most men began work within days of arriving in Australia.

Ultimately, more than 100 000 people migrated from more than 30 countries, including Austria, Cyprus, Czechoslovakia, England, Estonia, Finland, France, Germany, Greece, Hungary, Ireland, Italy, Jordan, Norway, Poland, Portugal, Romania, Russia, Scotland, Sweden, Switzerland, Turkey, USA and Wales. Seven towns and more than 100 temporary camps were built throughout the Snowy Mountains to house them.

The work was hard in harsh conditions, most of it underground. More than 120 workers died to build the Snowy Mountains Scheme.

The Snowy Mountains Scheme consists of 16 dams, 7 power stations and 145 kilometres of tunnels.

Source: Go Facts Australia: *Migration*, p. 22, Blake Education

Research

After World War II, migrants came from all over Europe to work on the Snowy Mountains Scheme. What were some of the challenges the migrants faced in moving to a new country and living in temporary camps?

Questioning

What would you ask someone who had just arrived from another country to work on the Snowy Mountain Scheme? Write three questions.

1

2

3

Analysing

What does the text tell you about life in Australia after 1949? How did the Snowy Mountain Scheme change communities and the way that people lived?

Communicating

Create a poster to encourage migrants from Europe to choose to come to Australia to work and live. Use a powerful heading, dot points and appropriate images.

Evaluating and Reflecting

Once the Snowy Mountain Scheme was completed in 1974, most of the migrant workers stayed in Australia. Why do you think they did this instead of returning to Europe?

UNIT 7

An interview with my Grandad

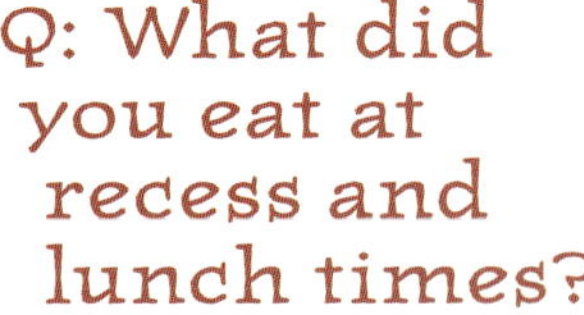

Q (Question): How old were you when you first went to school?

A (Answer): I was six years old when I was in Year 1. I went to a public school in Melbourne, Victoria. It was 1955.

Q: How did you get to school?

A: Everyone walked to school in those days. I would call in at my friend's house and we would walk together. There were no school crossings – we had to cross the roads by ourselves.

Q: How many children were in your Year 1 class?

A: There were 50 of us! I still have my class photo.

Q: What did you use to write with?

A: I learned to write with a pencil. Then I was allowed to use a fountain pen with a nib. I had to fill it up with ink from a bottle and use blotting paper to dry it on the page. My father wrote with thin slate pencils on little black slates.

Q: What sorts of things did you learn?

A: When I was in Year 4, we had to learn our times tables and spellings by chanting them all together. We also wrote stories – this was called 'composition'. We learned poems off by heart. I can still remember some of them. My favourite was *The Ant Explorer*, by C.J. Dennis.

Q: What did you eat at recess and lunch times?

A: At recess, or 'little lunch', we all drank a small bottle of milk. It was free and was delivered to the school every day. Sometimes it was left out in the sun and was horrible because it was warm and sour. Our school had a tuckshop where you could buy sandwiches, pies and sausage rolls. They also sold cream buns, soft drinks and lollies. I usually walked home for lunch because my mum was home and we lived near the school.

Q: What happened if you did something wrong?

A: Once I threw a block of wood up in the air and it hit a girl on the head, making her bleed. I got four hits of the cane for that! The Headmaster would tell you to hold your hand out with the palm up and then hit your hand hard with the wooden cane. It hurt like mad. My hand was so sore I couldn't write for the rest of the day.

Q: What were your school desks like?

A: Our desk was quite long, so eight of us could sit there. We sat on a long wooden bench, as long as the desk. We didn't have chairs.

Source: Australian History Centres: *Lower Primary*, p. 9, Blake Education

TARGETING HASS 3 © PASCAL PRESS ISBN: 9781925726046

Research

The Australian Bureau of Statistics (ABS) collects data about people, their families, work and culture. Visit their website at *https://www.abs.gov.au*
Record three things that you found interesting.

1 ______________________________

____

2 ______________________________

____

3 ______________________________

____

Analysing

How is your school life similar to and different from Grandad's school life?

____

____

____

____

____

____

When would you prefer to go to school – then or now? Why?

____

____

Questioning

In the interview, the boy asks his Grandad lots of interesting questions about his life at school. Your task is to interview one of your parents about what life was like when they were children. You could ask about the games they played, the groups they joined, and the activities they participated in.
Write at least three questions.

____

____

____

____

____

____

____

____

____

____

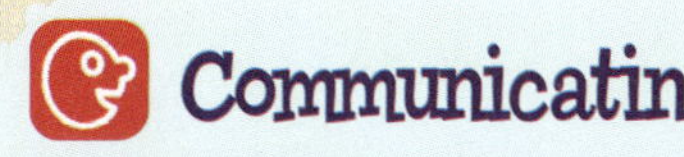

Communicating

Collect pictures, photographs and objects to represent your life at school. Use them to create a collage.

Evaluating and Reflecting

In Grandad's time, he would get free milk and he could buy cream buns, soft drinks and lollies from the canteen. How has the food at school changed over time? Have those changes been positive or negative? Why?

____

____

____

UNIT 8

HISTORY

Victor Chang (1936–1991)

Victor Chang was one of Australia's most talented and pioneering surgeons.

Victor was born in China to Australian-born Chinese parents. He decided to become a doctor after his mother died of cancer when he was 12 years old. Victor moved to Australia when he was 15. He studied medicine at The University of Sydney. He then developed his surgery skills in England and the United States.

In 1972, Victor returned to Australia to work at St Vincent's Hospital in Sydney. He worked alongside Dr Harry Windsor, who had performed Australia's first heart transplant in 1968. In 1980, when new drugs increased the number of successful heart transplants, Victor set up a transplant program at St Vincent's. From 1984–90, his surgical team completed more than 197 heart transplants and 14 combined heart-lung transplants. Of these patients, 90 per cent lived longer than a year after their operations.

Victor became concerned about the shortage of organ donors. With a team of scientists and engineers, he helped to develop an artificial valve, to control the flow of blood in and out of the heart. But before his next innovation – an artificial heart – was completed, Victor was tragically murdered.

In 2000, Victor Chang was voted Australian of the Century by the people of Australia. The Victor Chang Cardiac Research Institute continues his work, as it studies the causes and treatment of heart disease.

Victor worked to develop an artificial heart.

Did you know?
More than 10 000 Australians die from heart failure every year.

Source: Go Facts Australia: *Great Australians*, p. 26, Blake Education

TARGETING HASS 3 © PASCAL PRESS ISBN: 9781925726046

What is the nature of the contribution made by different groups and individuals in the community?

Research

What is heart disease? How is it caused? How can it be prevented?

Analysing

Why do you think Victor Chang was voted Australian of the Century?

Questioning

Write four questions that are answered in the information report about Victor Chang. Include different types of questions: when, how, what, why or who.

1

2

3

4

Communicating

Using your research about heart disease, write a one-minute speech to convince your audience they should live a healthy lifestyle. Present your speech to the class.

Evaluating and Reflecting

What characteristics do you think are important in an 'Australian of the Year'? What qualities should they have? Why would someone be given this award?

HISTORY

Ash Barty (1996-

Ashleigh Barty was Australia's – and the world's – best female tennis player in 2019.

Ashleigh, commonly known as 'Ash', was born in Ipswich, Queensland. Her father is a descendent of the Ngarigo people from the alpine regions of New South Wales and Victoria. Her mother's family migrated to Australia from England.

Growing up in Ipswich, sport played an important part in Ash's life. Along with her two older sisters, she played tennis and netball.

Ash started training with her long-term tennis coach in Brisbane when she was four years old. She had excellent hand-eye coordination and, even then, she was very determined. Ash would practise at home by hitting a tennis ball against the wall of her house. By the age of 12, she was playing against adults.

Did you know?
Ash Barty's hobbies are fishing, reading and playing computer games.

As a junior tennis player, Ash rose to be Number 2 in the world. In 2011, she won the junior Wimbledon championship when she was just 14. Two years later, she was runner-up in three Grand Slam doubles competitions, one of which was the Australian Open.

But Ash found the travelling difficult. She spent a lot of time in Europe, away from Australia and her family. She wanted to be more like a normal teenager. So in 2014, Ash took a break from tennis. She started playing semi-professional cricket with the Brisbane Heat in the Women's Big Bash League. Ash's top score was 39.

Despite having some success in cricket, Ash decided to return to the game she loved – tennis – in 2016. Since then, her tennis rankings have risen quickly. In 2019, shortly after winning the French Open, Ash was named the world's Number 1 female tennis player. This is a feat that no Australian woman has achieved since Evonne Goolagong Cawley in 1976.

But Ash is much more than a great tennis player. She is also the National Indigenous Tennis Ambassador. Her aim is to get more Indigenous children like herself to play tennis. Ash is proud of her Indigenous heritage and has been described as an "Inspirational example to all Australians" by Evonne Goolagong Cawley.

Photo: © Zhukorvsky, ID 142270893, Dreamstime.com

TARGETING HASS 3 © PASCAL PRESS ISBN: 9781925726046

Research

Who is Yvonne Goolagong Cawley?
What are some of her achievements?

____

____

____

____

____

____

____

____

____

Questioning

Are these statements **true** or **false**?

1 Ash played cricket as a child.

____

2 Ash did not enjoy all the travelling.

____

3 Ash's mother was born in Australia.

____

4 Ash is a very focussed player.

____

5 Ash has an Indigenous Australian background.

____

6 Ash does not represent Indigenous Australians.

____

Analysing

What has Ash Barty contributed to the community?
How has this contribution been important?

____

____

____

____

____

____

____

Communicating

Make a poster to encourage people to 'give back' and contribute to the wider community by:

- helping others
- donating to local charities
- volunteering at community events
- participating in cultural activities.

Evaluating and Reflecting

Do you think Ash Barty should be a role model for Australian children? Why do you think this?

____

____

____

____

UNIT 10

Symbols and celebrations

Symbols and celebrations remind us of what is special about Australia.

National celebrations are a chance for people to come together at events that are important to them. People celebrate special days and honour people who have died.

Symbols are also important. When we wear the national colours – green and gold – we show we belong together. The colours come from Australia's floral emblem, the golden wattle. Our national gemstone is the opal, known for its brilliant colours. It is mined in the Australian outback. In Aboriginal stories, a rainbow created the colours of the opal when it touched the earth.

The Commonwealth Coat of Arms. The symbols on the shield represent (top row, left to right) NSW, Victoria, Queensland; (bottom row, left to right) South Australia, Western Australia and Tasmania.

The Commonwealth of Australia uses a coat of arms to show its authority and ownership. The coat of arms has a shield with the badge of each Australian state. The shield is held by native animals, the kangaroo and the emu. The border of the shield symbolises that the states are part of a federation. Above the shield is the seven-pointed Commonwealth Star. It has a point for each of the six states, and one point for the Australian territories. Behind the shield is golden wattle.

Did you know?
In an earlier version of the Commonwealth Coat of Arms, the emu held one leg up against the shield.

Source: Go Facts Australia: *Symbols and Celebrations*, pp. 4–5, Blake Education

HISTORY

TARGETING HASS 3 © PASCAL PRESS ISBN: 9781925726046

Research

The Australian flag is an important national symbol. It features the Union Jack, the Southern Cross and the Federation star. Why are these features on the flag? What do they represent?

____
____
____
____
____
____
____
____
____

Questioning

How does the coat of arms represent the whole country?

____
____
____

How does the coat of arms represent the different states?

____
____
____

How does the coat of arms represent our long history of Aboriginal culture?

____
____
____

Analysing

Why are national symbols, such as our coat of arms and flag, important to different groups and individuals?

____
____
____
____
____
____

Communicating

Create a timeline showing when:

- the coat of arms was created
- the national anthem was written
- the national flag was designed
- an Australian sporting team first wore green and gold
- Australia became a nation.

Evaluating and Reflecting

If you were to create a new national symbol for Australia, what would you choose? Why?

____
____
____
____

UNIT 11

Sorry Day and Harmony Day

HISTORY

Sorry Day is for remembering past mistakes. Harmony Day is for celebrating the Australia of today.

Sorry Day is 26 May. It is a day to remember that the British treated Aboriginal and Torres Strait Islander people badly when they came to Australia. It is also a day to recognise the strength of the people who survived.

The British took land belonging to Aboriginal and Torres Strait Islander people. They killed or hurt the people they found – sometimes on purpose, sometimes by accident. In the years that followed, the Australian Government removed many Aboriginal and Torres Strait Islander children from their families. They wanted to raise them like British children. Families and communities were separated for years. Some people never returned to their families. They became known as the Stolen Generations. They told their stories in the *Bringing Them Home* report in 1997 – the first Sorry Day was the following year.

Being sorry about the past can be the first step for a better future.

Harmony Day celebrates the fact that almost half of Australians were born overseas or have a parent who was.

Did you know?

The message of Harmony Day is "Everyone Belongs".

Harmony Day is a chance to celebrate all the nationalities that make up Australia. It is on 21 March. On Harmony Day, people share their culture through dance, art, film, sport, music and cooking. This allows people to learn about other cultures and teach people about their own.

Source: Go Facts Australia: *Symbols and Celebrations*, p. 14, Blake Education

TARGETING HASS 3 © PASCAL PRESS ISBN: 9781925726046

What is the nature of the contribution made by different groups and individuals in the community?

UNIT 11

Research

Harmony Day is about sharing culture. Research a culture other than your own. Write five interesting facts you found out.

Questioning

Imagine a new student comes into your class. They have recently arrived from another country. What could you ask them to get to know them better? Write three questions.

1

2

3

Analysing

What mistakes were made by the Australian Government? Why was it important for the Prime Minister to say 'Sorry' for what happened?

Communicating

Create a word cloud to represent Harmony Day and its message **Everyone belongs**. Include at least 10 different words.

Evaluating and Reflecting

Why is it important for everyone to celebrate Harmony Day? What would Australia be like if we didn't have Harmony Day?

Vietnamese New Year (Tet)

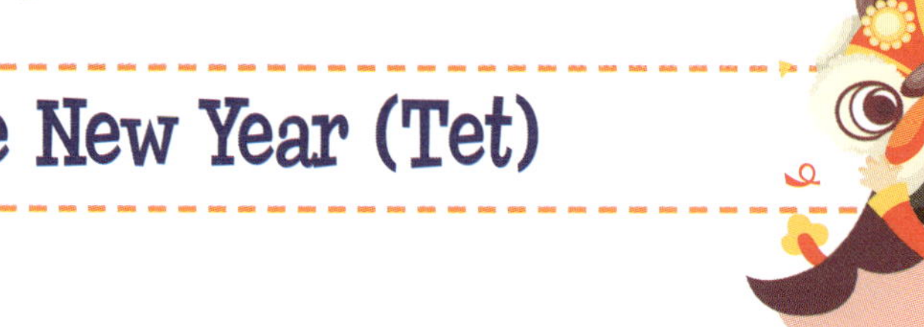

Vietnamese New Year is the most important celebration of the year for Vietnamese people. It is also called Tet.

People clean their houses and cook special food like rice cakes.

Children are given lucky money in bright red envelopes.

They visit the homes of their relatives.

It is celebrated on the first day of a special calendar called the lunar calendar. This is the beginning of spring.

People say special prayers and light candles at temples.

They bang drums, bells and gongs to scare away evil spirits.

There is a big street parade. People wear lion costumes and do a 'lion dance' for good luck.

Source: Australian History Centres: *Lower Primary*, p. 17, Blake Education

How and why do people choose to remember significant events of the past?

Research

Find out about Chinese New Year.
How is this celebration similar to Vietnamese New Year? How is it different?

Similarities	Differences

Questioning

What special events does your family celebrate?

Why do you celebrate these events?

How do you and your family celebrate?

Analysing

Interview the people in your class. What special events do they celebrate? Tally their responses. Then make a column graph to show the results.

Event	Tally	Total
Birthday		
Name day		
Christmas		

Communicating

Discuss the results of your survey with your friends. What does your graph tell you about the events people celebrate? Talk about why people celebrate these types of events.

Evaluating and Reflecting

What is your most important celebration? Why is it so important to you?

UNIT 13

Ramadan

HISTORY

Ramadan is a time for Muslims to fast and pray. They try extra hard to live God's way.

Did you know?

Young children, pregnant women, and people who are ill don't have to fast during Ramadan.

Ramadan lasts for 29 or 30 days.
It begins when a crescent moon appears in the sky. During this time, Muslim people get up before sunrise to eat a meal called *suhoor*. They don't eat or drink at all during the day. After dark, they eat a meal called *iftar*.

Muslims pray each day during Ramadan. They share their money with the poor to show their thanks to God.

After Ramadan is a celebration called *Eid-ul-Fitr*. It means Festival of Breaking the Fast. Muslims can eat again during the day. To get ready for Eid-ul-Fitr, people put up lights and decorations. They cook lots of food to share with visitors. On the morning of the festival, people wake early. They dress in new clothes. They go to the mosque for prayers. Finally, they gather with family and friends to enjoy a meal.

Many Muslims read the whole Koran, the holy book of Islam, during Ramadan.

Ramadan is one of the Five Pillars of Islam, the acts that all Muslims must follow.

You can wish a Muslim a happy Ramadan by saying "Ramadan Mubarak". It means, "Have a blessed Ramadan".

Source: Go Facts Australia: *Special Days*, pp. 12–13, Blake Education

TARGETING HASS 3 © PASCAL PRESS ISBN: 9781925726046

Research

Ramadan is a religious celebration. Find out about another religion and one of its special days.

Religion:

What do they celebrate? When does it happen? Why do they celebrate this?

Questioning

1 When does Ramadan start?

2 What meal do Muslims eat before sunrise during Ramadan?

3 Name two ways that Muslims show their thanks during Ramadan

4 What does Eid-ul-Fitr mean?

5 What is the name of the Muslim place of worship?

6 What is the Koran?

Analysing

Young children, pregnant women and people who are sick don't have to fast during Ramadan. What do you think is the reason for this?

Communicating

Create an acrostic poem using the word **Ramadan**.

Evaluating and Reflecting

What would be the most challenging part of Ramadan for you? Why do you think that?

Anzac Day

HISTORY

A recruitment poster from 1915 encouraging men to enlist in the war

World War 1 (1914-1918)

When Britain went to war with Germany in August 1914, most Australians still thought that they were more British than Australian. Many men rushed to enlist because they felt they had to stand by the 'mother country'. Some men believed that it was their duty, while others thought it was a chance for excitement and travel. They all believed the war would be over quickly.

The Australian War Memorial, Canberra. It combines a museum, shrine and archives of Australia's war records.

ANZAC

When soldiers from Australia and New Zealand arrived in Europe to fight in the war, the British generals decided they needed a name. They became the Australian and New Zealand Army Corps, or the Anzacs.

During World War I, the Ottoman Empire (now Turkey) had an agreement (alliance) with the Germans. So the Allied powers of Britain and France decided that they would attack the Turks. On 25 April 1915, the Anzacs began to land troops on the Gallipoli peninsula, at what is now known as Anzac Cove. But the landing did not go to plan. The boats did not land at the right place. Steep, rocky cliffs overlooked the beach, giving the Turkish soldiers a clear advantage. As a result, the Anzac troops couldn't break through the Turkish defences. Despite this, their commander, Sir Ian Hamilton, told them to, "dig, dig, dig, until you are safe".

By the time the Anzac troops were evacuated from Gallipoli in December 1915, more than 8000 Australian soldiers had died.

Lone Pine

One of the most memorable battles by the Anzacs at Gallipoli was at Lone Pine. It was named for the single pine tree that marked the battle site. On 6 August 1915, the Australians captured the main Ottoman trench. But four days of hand-to-hand combat resulted in over 2000 Australian casualties. The Lone Pine Cemetery and Lone Pine Memorial at Gallipoli commemorate the soldiers who died there.

Anzac Day

Every year on 25 April, we remember the anniversary of the Anzac landing at Gallipoli. Through Anzac Day, Australians commemorate "the contribution and suffering of all those who have served". (Australian War Memorial website). It is our national day of remembrance.

Lest We Forget.

Anzac memorial on Anzac Bridge, Sydney

Source: *Blake's Australian History Guide*, p. 82, p. 86, Pascal Press

TARGETING HASS 3 © PASCAL PRESS ISBN: 9781925726046

How and why do people choose to remember significant events of the past?

Research

What do red poppies represent?

Analysing

Look at the recruitment poster from 1915 in the text. Do you think it is effective in persuading men to enlist? Explain your answer.

Questioning

Imagine you could talk to an Anzac soldier who fought at Gallipoli. What would you ask them about their experiences? Write three questions.

1

2

3

Communicating

Charles Bean was an Australian reporter who landed on the beaches of Gallipoli on 25 April 1915. Write your own short newspaper report about the landing.

Evaluating and Reflecting

The original Anzac Day happened over 100 years ago. Why do we still remember it? Why is it important to remember events from the past?

Australian Landscapes

Australia has many different landscapes.

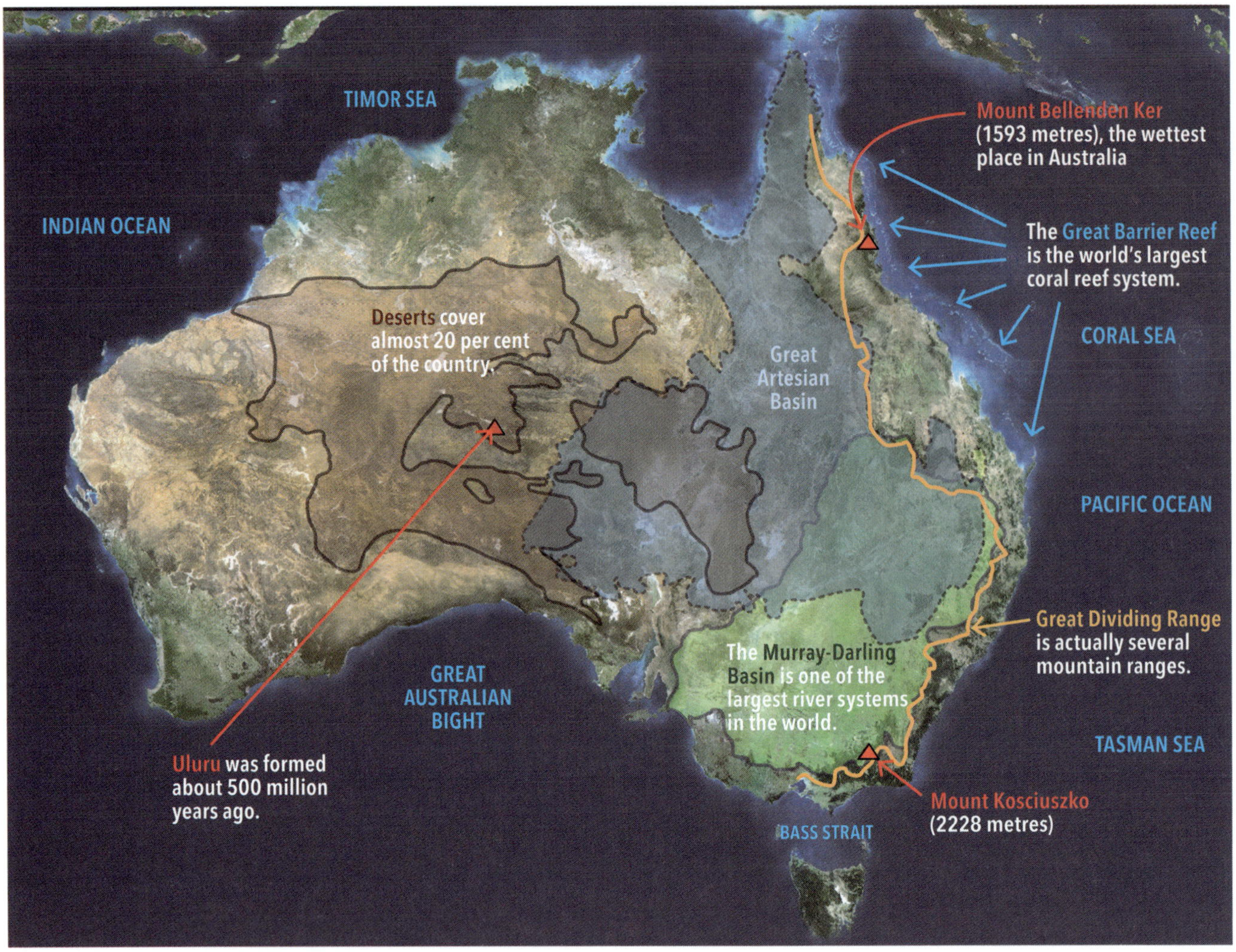

TIMOR SEA

INDIAN OCEAN

Mount Bellenden Ker (1593 metres), the wettest place in Australia

The Great Barrier Reef is the world's largest coral reef system.

CORAL SEA

Deserts cover almost 20 per cent of the country.

Great Artesian Basin

PACIFIC OCEAN

Great Dividing Range is actually several mountain ranges.

The Murray-Darling Basin is one of the largest river systems in the world.

GREAT AUSTRALIAN BIGHT

TASMAN SEA

Uluru was formed about 500 million years ago.

Mount Kosciuszko (2228 metres)

BASS STRAIT

Australia is the lowest, flattest and driest inhabited continent. There are tropical rainforests in the north, deserts in the centre, and temperate regions in the south.

Around 150 million years ago, the shallow Eromanga Sea covered nearly one-third of the continent. Now, about 70 per cent of Australia is arid or semi-arid. Most of the rain in the north of the country falls in the tropical summer. The south gets most of its rain in the colder months.

Australia's mountains play a role in where rain falls. The Great Dividing Range forces moist air to rise, which causes rain along the east coast. Western Tasmania is very wet because its mountains cause rain to fall. Central Australia has more rainfall than other deserts of the world because there are no mountains in the continent's west to stop moist air travelling in from the Indian Ocean.

Australians have changed the landscape to benefit from it. They dammed the Murray, Darling and Murrumbidgee Rivers to control their flow. Reservoirs, dams and lakes hold water used to grow more than one-third of Australia's food. Water for farming also comes from the Great Artesian Basin, one of the world's largest underground reservoirs. It lies beneath 22 per cent of Australia.

Source: Go Facts Geography: *Australia*, pp. 16–19, Blake Education

GEOGRAPHY

TARGETING HASS 3 © PASCAL PRESS ISBN: 9781925726046

Research

Lake Eyre in South Australia is the lowest point in Australia, at 15 metres below sea level. Investigate why it is a significant natural feature.

Questioning

What is a temperate environment?

What is an arid environment?

What is a reservoir?

How high is Mt Kosciuszko?

Which states does the Great Diving Range cross through?

Why is the Great Barrier Reef special?

Analysing

Look at the map from the text.
Uluru is sometimes called the 'heart' of Australia. Suggest reasons for this.

Communicating

Create a map of your local area.
Show the special natural features (rivers, beaches, mountains, national parks) and the special human features (buildings, bridges, parks).

Evaluating and Reflecting

Choose one of the natural or human features from your map of your local area.
Why is it special? How do you interact with this feature?

UNIT 16

States and Territories

1788

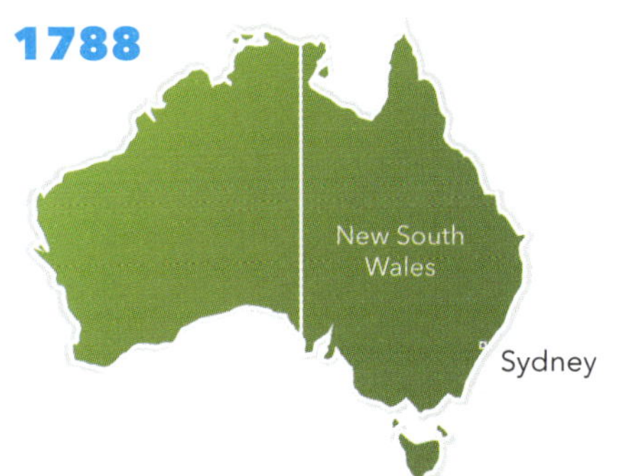

1825

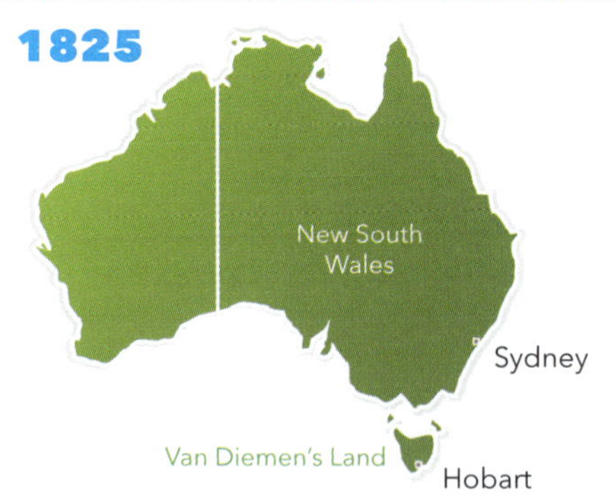

1836

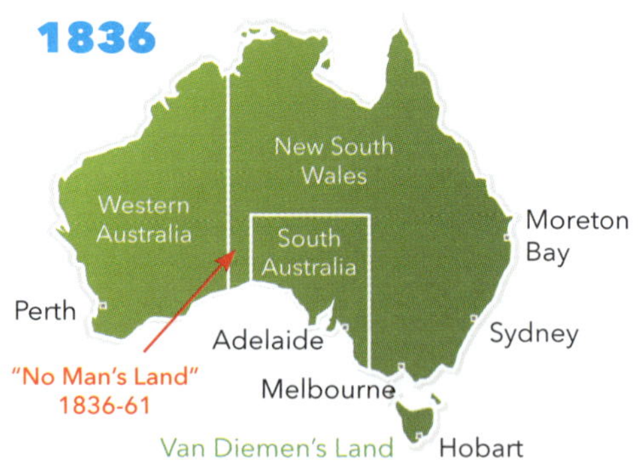

1856

1859

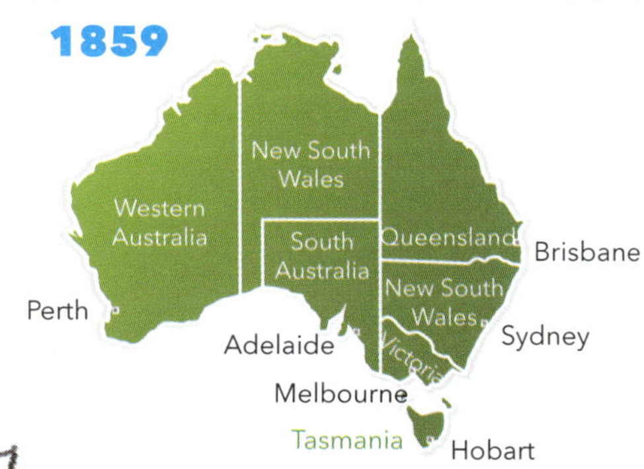

1863

Northern Territory
NT separated from SA in 1911
Western Australia
Queensland
South Australia
Brisbane
New South Wales
Perth
Adelaide
Victoria
Sydney
Melbourne
Tasmania
Hobart

The British colonies in Australia grew to become states and territories in a federation.

Originally, James Cook only claimed the eastern half of the country for Great Britain. The Dutch did not claim the western half - New Holland - as theirs. By the early 1800s, England feared that France, its enemy, might colonise the western half. It decided to start a new colony in the west. In 1829, Captain Charles Fremantle claimed New Holland for Great Britain.

Colonial governors and surveyors chose sites for the colonies. They made their decisions based on the need for safe harbours, fresh water and suitable land for farming. The Secretary of State for the Colonies, a politician in England, decided where to draw the borders between colonies. He advised Queen Victoria, who made the borders law. Most borders were straight lines. But people in New South Wales and Victoria argued whether their border should be the Murray River or the Murrumbidgee River. In 1842, Queen Victoria decided it should be along the Murray River. In 1901, she declared Australia a federation. The colonies became states as part of one nation.

Did you know?

Australia's shortest border is between Tasmania and Victoria. Boundary Islet in the Bass Strait – just 85 metres wide – is the only land on the line of latitude that separates the two states.

Source: Go Facts Geography: *Australia*, pp. 10–11, Blake Education

GEOGRAPHY

TARGETING HASS 3 © PASCAL PRESS ISBN: 9781925726046

Research

Investigate the state you live in.
Find out these things.

1 Coat of arms

2 Floral emblem

3 Animal emblem

4 Bird emblem

5 Fish emblem

6 Gemstone

7 Colours

8 Motto (Not all states and territories have one.)

Questioning

Why do you think the Dutch did not claim New Holland, the western half of Australia?

Communicating

Create a timeline to show the development of Australia's states and territories from a British colony in 1788 to the present day. Remember to include federation and the Australian Capital Territory. Hint: use a scale of 1 cm = 2 years.

Analysing

Look at a map of Australia. What do you notice about the locations of the capital cities in each state? Why do you think they are located where they are?

Evaluating and Reflecting

Was Queen Victoria's decision to use the Murray River as the border between New South Wales and Victoria a good decision? What could be some reasons for and against this decision?

UNIT 17

Lost Cities of Stone

When Dutch sailors first saw the pillars of the Nambung National Park from their ships, they thought they were looking at an ancient city. Australia has many rock formations like this. Three rock formations in the Northern Territory are known as the Lost Cities – in Watarrka National Park, Litchfield National Park and Limmen Bight.

The columns of the Lost City in Litchfield National Park, Northern Territory, form a maze of rocky alleyways of 500-million-year-old sandstone.

Lost Cities

The columns in the Northern Territory's three lost cities are made of sandstone. The water seeping down through the joints in the sandstone rock have cut out columns, leaving pillars of free-standing stone. These look like the abandoned buildings of a crumbling, long-ruined civilisation.

Thousands of round-domed columns in the Southern Lost City near Cape Crawford, Limmen Bight. Aborigines used this sacred place to perform initiation ceremonies.

The stone cities of Watarrka and Litchfield can be explored by walking between the pillars. But the Southern Lost City and Western Lost City of Limmen Bight, about 250 km west of the Queensland border with the Northern Territory, are usually seen from the air. They consist of exposed crimson pillars that are up to 25 m high and are made up of 95% silica with a rust-red iron crust.

The remote and isolated Lost City near the Tawallah Range in Limmen National Park, Northern Territory, is best viewed from the air.

the FACTS!

THE SANDSTONE STRUCTURES of the Lost City of Watarrka National Park are more rounded than those seen in other Australian landforms of the same name.

THE PILLARS in the Southern Lost City in Limmen National Park are made up of extremely old rocks, some of which are 1.4 billion years old.

THE WESTERN LOST CITY is far more remote and is accessible only by four-wheel-drive vehicle, so it is usually explored from the air.

LITCHFIELD NATIONAL PARK was first explored by Europeans in 1864. Before then it had been home to the Wagiat Aboriginal people for many thousands of years.

Watarrka's Lost City is made up of more-rounded, beehive-shaped domes.

Source: Amazing Facts about Australia: *Iconic Landscapes*, p. 35, Steve Parish

GEOGRAPHY

TARGETING HASS 3 © PASCAL PRESS ISBN: 9781925726046

Research

There is a similar stone formation in Western Australia known as *The Pinnacles*. Find out four interesting things about these stone pillars.

1 ______________________________

____

2 ______________________________

____

3 ______________________________

____

4 ______________________________

____

Questioning

Imagine you could talk to a geologist (earth science expert) about the pillars of Nambung National Park.
Write four questions you would ask.

1 ______________________________

____

2 ______________________________

____

3 ______________________________

____

4 ______________________________

____

Analysing

Why is it best to view the Lost City near the Tawallah Range in Limmen National Park from the air? What would be some of the challenges if you tried to see it from the ground?

____

____

____

____

____

Communicating

Imagine you are a tour guide working in the Northern Territory.

What you would say to a tour group visiting one of the Lost Cities of Stone? What important things could you tell them about caring for this natural feature?

Write a short speech you could give to the tour group.

Evaluating and Reflecting

Different groups, like tourists, scientists and local Aboriginal peoples, would all see the Lost Cities of Stone in different ways. What effect would these different views have on protecting these natural features?

____

____

____

____

UNIT 18

Old Government House, Parramatta

Photograph, 2016

This is Old Government House in Parramatta, New South Wales. It was built in 1799 and is Australia's oldest public building. This was the home of each new governor who was sent out to Australia from England. The main part of the house was built by Governor John Hunter. Other parts were added as time went on. Artists have painted pictures and taken photos of Old Government House over the years. These pictures show that the building hasn't changed very much, even though it is more than two hundred years old.

Today, Old Government House is a World Heritage site. This means that it is protected by certain laws and managed by people who will take care of it properly. It needs to be preserved because it tells us about the history of the early British people who came to Australia in the 1700s.

The Burramatta people are the traditional owners of the Darug land on which Old Government House is built.

Many sites important to Indigenous Australians are also listed as World Heritage sites. This means they are protected for future generations of Australians to visit and learn about.

Painting, 1805

Engraving, 1819

Photograph, 1932

Source: Australian Geography Centres: *Lower Primary*, p. 5, Blake Education

GEOGRAPHY

Research

Name the ten governors who lived in Old Government House. When did they serve?

Governor	Dates served

Questioning

Create a short quiz about Old Government House for your classmates. Challenge them to complete it. Write five questions.
Hint: Use different question words: when, how, what, why and who.

1 ______________________________

____

2 ______________________________

____

3 ______________________________

____

4 ______________________________

____

5 ______________________________

____

Analysing

Why are some features chosen as World Heritage sites? What does this mean?

____

____

____

____

____

____

Communicating

Create a welcome sign for visitors to Old Government House. What important information do you need to tell them? What rules should they follow?

Evaluating and Reflecting

Name a human feature in the area where you live that should be preserved for future generations. Give reasons for your choice.

____

____

____

UNIT 19

Australia's Neighbours in the Asia-Pacific

GEOGRAPHY

Australia's neighbours are in an area we call the Asia-Pacific region. These are countries that are found in or near the western part of the Pacific Ocean.

A compass is something used to figure out directions. On a map, a compass shows you the direction where places are. A compass always points North, which is where the top of the Earth is.

A compass has four main points: North, South, East and West. In-between these are half points: North-East, North-West, South-East and South-West.

Here are some countries in the Asia-Pacific and the directions they are from Australia.

DIRECTION	COUNTRY
South-East	New Zealand
North	Papua New Guinea
North	Philippines
North	Japan
North-West	Malaysia
North-West	Thailand
North-West	China
North-East	Vanuatu

Source: Australian Geography Centres: *Lower Primary*, p. 25, Blake Education

Research

Choose one of Australia's Asia-Pacific neighbours. Find out these things.

Country: ______________________

1 Population:

____

2 Languages:

____

____

3 Climate:

____

____

4 Landmarks (natural and constructed):

____

____

____

____

5 Native flora (plants):

____

____

6 Native fauna (animals):

____

____

Analysing

What do you notice about many of the countries in the Asia-Pacific region between Australia and the continent of Asia? Is this important? Why?

____

____

____

____

____

Questioning

1 Which country is north of Australia and east of China?

____

2 Which country is south of New Caledonia and south-east of Indonesia?

____

3 Which country is west of the Philippines and east of Thailand?

____

4 Which country is north-west of Malaysia and south-west of Japan?

____

5 Which country is east of Indonesia and north of Australia?

____

Communicating

Imagine you are going on a holiday to one of Australia's neighbours in the Asia-Pacific region. Which country would you visit? Why?

Evaluating and Reflecting

ASEAN (Association of South-East Asian Nations) has the motto: **One Vision, One Identity, One Community**. ASEAN was set up to develop cooperation between the member countries. Why is this important for the countries in the Asia-Pacific region?

____

____

____

____

____

More than One Asia

Comparing two countries in Asia shows just how diverse the continent is.

Japan is a series of islands with rugged, mountainous terrain. Its climate is tropical in the south, ranging to cool temperate in the north.

Japan is a wealthy, industrialised country. Following its defeat in World War II, Japan grew to become the third largest economy in the world. It is one of the most advanced producers of cars, electronic equipment and tools. Japan's population of 127 million people mainly lives in urban areas. It is a homogeneous society - 98.5 per cent are of Japanese descent - but it is ageing. One-quarter of the population is 65 years or older.

Nine out of 10 Japanese people live in urban areas.

Did you know?
The capital of Japan, Tokyo, isn't just the largest city in the world – it's the largest in the history of the world.

Japanese women are having fewer children. The country's population is falling.

Cambodia is half the size of Japan and has a small fraction of its population (15 million). Most Cambodians live in rural areas. The country has a short coastline and is mostly low plains. It has a tropical climate.

Cambodia is one of the least developed countries in the world. It is still recovering from a civil war in the 1970s, in which almost 2 million Cambodians died and the economy was destroyed. Its GDP is less than one-hundredth the size of Japan's. Half of its budget is donated by other countries. Cambodia's economy is growing, driven by the garment, construction, agriculture and tourism industries.

Like Japan, Cambodia has a homogeneous society. Nine out of 10 people are of Khmer descent. Unlike Japan, it is a young society - half the population is 25 years old or younger.

More than 2 million tourists visit Cambodia's ancient Angkor Wat every year.

One in four Cambodians older than 15 cannot read and write.

Source: Go Facts Geography: *Asia*, pp. 10–11, Blake Education

TARGETING HASS 3 © PASCAL PRESS ISBN: 9781925726046

Research

Angkor Wat is an ancient temple complex in Cambodia. Find at least five interesting facts about it.

Analysing

Half of Cambodia's budget is donated by other countries. Do you think that Australia should donate money to Cambodia? Why or why not?

Questioning

Are these statements **true** or **false**?

1 Cambodia is a wealthy country.

2 Fewer babies are being born in Japan.

3 Japan is an island.

4 Most Japanese people live in rural areas.

5 Tokyo is the largest city in the world.

6 Japan is smaller than Cambodia.

7 Many Cambodians can't read and write.

8 Angkor Wat is popular with tourists.

Communicating

Create a tourist brochure for Angkor Wat in Cambodia.

Evaluating and Reflecting

What did you learn about life in Cambodia? Did this information encourage you to visit or live there? Explain why.

UNIT 21

Maori

Māori arrived in New Zealand from East Polynesia in the thirteenth century.

The Māori were warrior hunter-gatherers. Tribes on the North Island also farmed vegetables. The population may have reached 100 000 before the British arrived.

British people colonised the country in the late eighteenth century. In 1840, the British Governor, on behalf of Great Britain, signed the Treaty of Waitangi with Māori chiefs. The treaty allowed Great Britain to colonise New Zealand in exchange for the Māori becoming British citizens, and keeping their land and tribal possessions.

Māori war dances (haka) include facial expressions (pūkana) that show the performers' passion. Tattoos (tā moko) show a person's ancestry and personal history.

Despite the Treaty, Māori and Pākehā (Māori word for white New Zealanders) fought over land ownership. There were several wars over the next 30 years. Today, the legal status of the Treaty of Waitangi still causes arguments.

A Māori greeting includes shaking hands and touching noses and foreheads (hongi).

Māori belong to iwi (tribes). Within each iwi are many hapū (clans or descent groups), each of which has one or more whānau (extended families). Māori are 15 per cent of New Zealand's population. In 1945, 26 per cent of Māori lived in towns and cities - now it is more than 80 per cent. One-quarter of Māori live near Auckland, New Zealand's largest city.

Māori have had a more positive experience of European colonisation than some first peoples, but they still experience disadvantage. On average, they earn less and their life expectancy is shorter than for Pākehā. About half of Māori leave high school without a qualification. More than 50 per cent of prisoners in New Zealand are Māori.

Did you know?

The Māori word for New Zealand is Aotearoa – "land of the long white cloud".

Māori culture is known for its wood carving for buildings and canoes.

Source: Go Facts Geography: *First Peoples*, pp. 20–21, Blake Education

GEOGRAPHY

TARGETING HASS 3 © PASCAL PRESS ISBN: 9781925726046

Research

The Māori people are known for the haka. What is it? How is it performed? What does it mean? Find out at least five interesting facts about the haka.

Questioning

Write five questions you have about New Zealand and the Māori people.

1 ______

2 ______

3 ______

4 ______

5 ______

Analysing

What do you think is the purpose of the wooden carvings on Māori buildings and canoes?

Communicating

A treaty is an agreement between two groups. It covers what they are allowed to do and what they are responsible for.

Create a treaty agreement between the students and teachers at your school. Share it with your class. Do they agree to the terms of the treaty? Does your teacher?

Evaluating and Reflecting

The Māori had a more positive experience of European colonisation than some first peoples, such as Australian Aborigines. Suggest reasons for this.

UNIT 22

My Asia

My name is Kenji Uranishi.

I was born and raised in Nara, Japan, but I've lived in Brisbane since 2004. Nara is the ancient capital of Japan and it is a beautiful place, surrounded by mountains. My mother still lives in the home where I grew up. When I visit her, I walk the streets where I used to play. I have wonderful memories of my childhood. I love mountains and they were right on my doorstep. In Brisbane there are not so many mountains nearby - I miss that.

Me and my family

The vegetable patch just behind the home where I grew up.

My grandmother

I work as a ceramicist.

When I arrived in Australia, I couldn't believe how expansive and deep blue the sky was. I was also surprised that Australia was so multicultural - there were people from all over the world, sharing their cultures. The scale of multiculturalism felt very different to that in Japan. I was also surprised at how many Australians were fascinated by Japan, loved its food and wanted to travel there. It made me feel proud!

One of the main differences between Japan and Australia is space. Australia is about 22 times bigger than Japan, yet Japan's population is far greater. You really notice how wide and open Australia is. Also, I noticed that Australian cakes are very sweet compared to Japanese cakes - I think Aussies have a sweet tooth!

I think one of the most interesting things for people to understand about Asia is that it's made up of so many different countries and cultures, languages and people. There are amazing things to learn and experience!

Source: Go Facts Geography: *Asia*, pp. 28–29, Blake Education

GEOGRAPHY

TARGETING HASS 3 © PASCAL PRESS ISBN: 9781925726046

Research

Capital city of Japan: ____________________

Flag of Japan:
(draw it here)

Japan's four main islands:

1 ____________ 2 ____________

3 ____________ 4 ____________

Japanese currency: ____________________

Highest mountain: ____________________

Native animals:

1 ____________________

2 ____________________

3 ____________________

Questioning

Write questions that have these answers.

Answer: Nara

Question:

____

____

Answer: Multiculturalism

Question:

____

____

Answer: Sushi

Question:

____

____

Answer: Ceramicist

Question:

____

____

Analysing

How is your life in Australia similar to and different from Kenji's life in Japan?

Similarities	Differences

Communicating

Imagine that you are Kenji. Write a postcard to your mother in Japan, telling her about the new country you live in. Make sure you compare your new way of life with your old way of life.

Evaluating and Reflecting

What would it be like to move overseas and live in another country, away from your family? What do you think were some of the challenges faced by Kenji when he did this?

____

____

____

People, Places and Spaces

The world is made up of places.

These places are where we live and grow up. They are where we learn and where we play. A place can be a city or a cubby house. A place can be a mountain range or a cave. A place can be a home, a park or a school library.

Places have different features. They can have natural features, such as a river. They can have managed features, such as a garden. They can have built features, such as a railway track. Many places have a mixture of different types of features.

Places change over time. People change them. They clear trees from the land to plant crops and grow food, and they build houses and roads. Nature also changes places. Heavy rain washes soil away from farms and fierce storms damage buildings.

Did you know?
The world's oldest botanical garden is in Italy. Built in 1545, it has a round garden bed in the middle, which represents the world.

Places have different meanings for different people. A person's family, faith and community connect them to places. Indigenous Australians have a special connection to places. The places they belong to – their country – are the source of their identity and culture. Many school and official events begin with an acknowledgement of country. This ceremony respects the importance of place to Indigenous Australians.

Attention
Dog Guardians
Pick up after your dogs - Thank you.
Attention Dogs
Grrrr, bark, woof.
Good dog.

Public places have rules, so that everyone can use the place.

A public place is shared by everyone. Parks, beaches and footpaths are examples of public places. We behave differently in a public place (playground) compared to a private place (our own backyard). People speak softly in libraries and cinemas. They share the space on footpaths and in car parks. They queue if they have to wait for something. Public places have rules, so that everyone can use the place safely and equally.

Commonwealth Park is on the banks of Lake Burley Griffin in Canberra. People go there to walk, play or relax. The gardens and grass are managed features. The sculptures, paths and toilets are built features. More than 400 000 people visit the Commonwealth Park Spring Festival every year.

People often name places. Indigenous Australians named their country. The Gadigal people fished in a cove they called Warrane. When the British arrived, they changed the name to Sydney Cove. Places can be named after famous people (Cooktown), what they look like (the Glass House Mountains), or the history of that place (Coal Point).

Source: Go Facts Geography: *People and Places*, pp. 4–5, pp. 12–13, Blake Education

GEOGRAPHY

TARGETING HASS 3 © PASCAL PRESS ISBN: 9781925726046

Research

Which town or suburb do you live in? Find out the origin of its name. Where did it come from? Who named it? What does the name mean?

____

____

____

____

____

____

____

____

Questioning

Write five questions that are answered in the text. Include different types of questions: when, how, what, why and who.

1 ____

____

2 ____

____

3 ____

____

4 ____

____

5 ____

____

Analysing

Choose a place near you. What meanings would it have for different groups of people? In what different ways would each group interact with this place?

Place:		
Group 1	**Group 2**	**Group 3**

Communicating

Signs in public places often use images so everyone can understand them.

Draw at least five symbols that represent places in your school or local community. For example: canteen, bike rack, office, drinking fountain.

Share these symbols with your friends. Do they understand what they mean?

Evaluating and Reflecting

Do we have too many rules in public places?
What is one rule that you think we don't need any more? Explain why.

____

____

____

UNIT 24

My Place

GEOGRAPHY

PERSONAL (MY HOME)

My dog Benji

Hi, my name is Jasmine! I'm seven years old. I live in a three-bedroom townhouse with my Mum, Dad and little brother Zach. I also have a dog named Benji. Our townhouse is inside a complex called Sunny Place.

Our house

LOCAL (MY TOWN, SUBURB OR DISTRICT)

I love living near the beach!

Our complex is in a suburb called Marina Bay on the Gold Coast. My school is very close to our house and Mum walks me there every morning. I like where we live because it's only a short drive to the beach! There are lots of places on the Gold Coast where I can go swimming or ride my bike.

Mum, Dad, Zach and me

REGIONAL (MY STATE)

The Gold Coast is in the state of Queensland. Queensland is in the north-eastern part of Australia. I was born in a hospital in Brisbane, which is about one hour north from the Gold Coast by car. Mum's parents, Nan and Pop, live on a farm in Coffs Harbour in New South Wales. We're taking a road trip to visit them during the next long weekend, as it takes a bit over four hours to drive there.

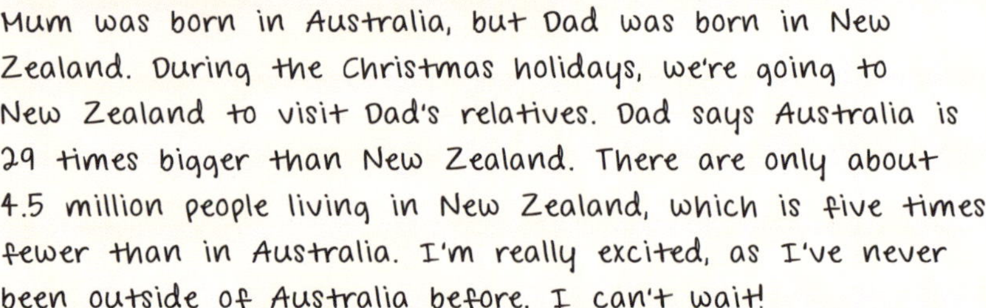

NATIONAL (MY COUNTRY)

Mum was born in Australia, but Dad was born in New Zealand. During the Christmas holidays, we're going to New Zealand to visit Dad's relatives. Dad says Australia is 29 times bigger than New Zealand. There are only about 4.5 million people living in New Zealand, which is five times fewer than in Australia. I'm really excited, as I've never been outside of Australia before. I can't wait!

Dad says New Zealand has lots of mountains – and sheep!

Source: Australian Geography Centres: *Lower Primary*, p. 31, Blake Education

TARGETING HASS 3 © PASCAL PRESS ISBN: 9781925726046

Research

Find out about the place where you live.

The type of home you live in: ____________________

____

Your home address: ____________________

____

Your suburb, city and state: ____________________

____

The places where you learn: ____________________

____

The places where you shop: ____________________

____

The places where you play: ____________________

____

Other important places for you: ____________________

____

Questioning

Write a series of questions to find out about the places that are important to someone else.

- Where ____________________

____

- What ____________________

____

- How ____________________

____

- Do ____________________

____

- ____________________

____

- ____________________

____

Analysing

Compare Jasmine's places with your own.

Place	Similarities to Jasmine's	Differences to Jasmine's
Home		
Relatives		
Recreation		
Holidays		

Communicating

Create a picture collage of your place. Include features that are personal, local regional and national.

Evaluating and Reflecting

What makes your place special? Do you like your place? Explain why.

____

____

____

UNIT 25

Places are similar and different

People like to know what the weather will be like each day. We have weather reporters to tell us if it will be sunny, or if we should expect rain. Knowing about the short-term weather and long-term climate of a place can help us decide what places to go to and when.

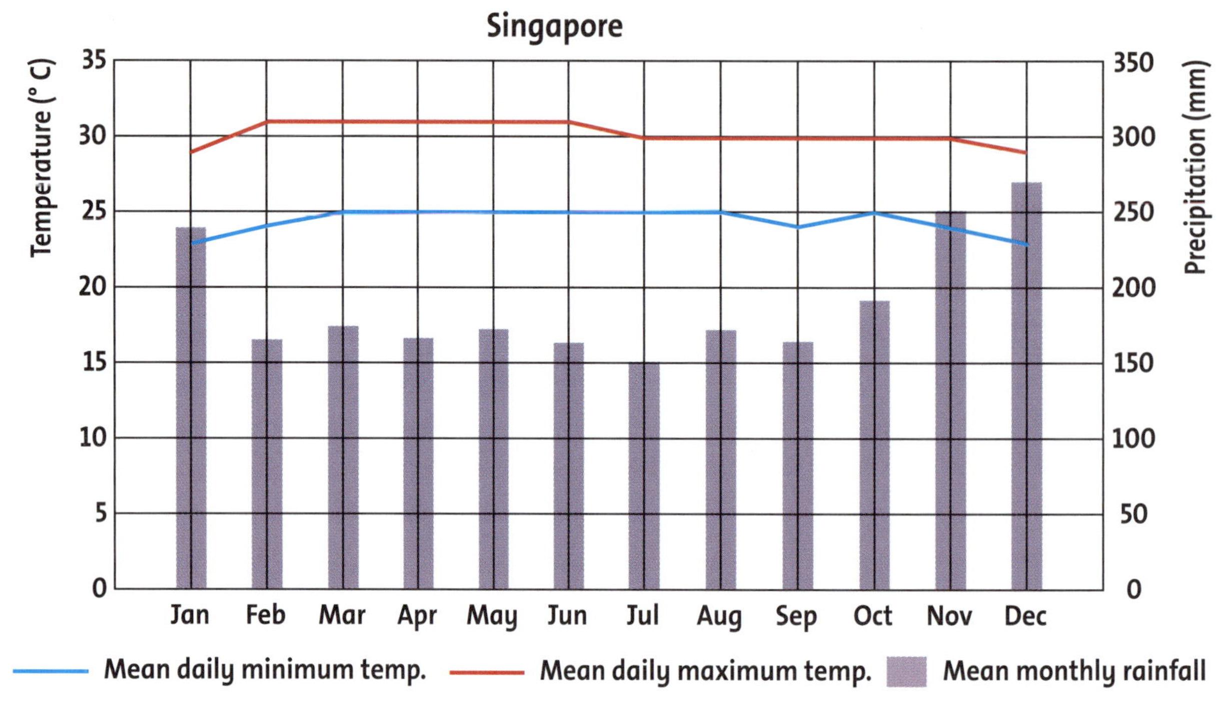

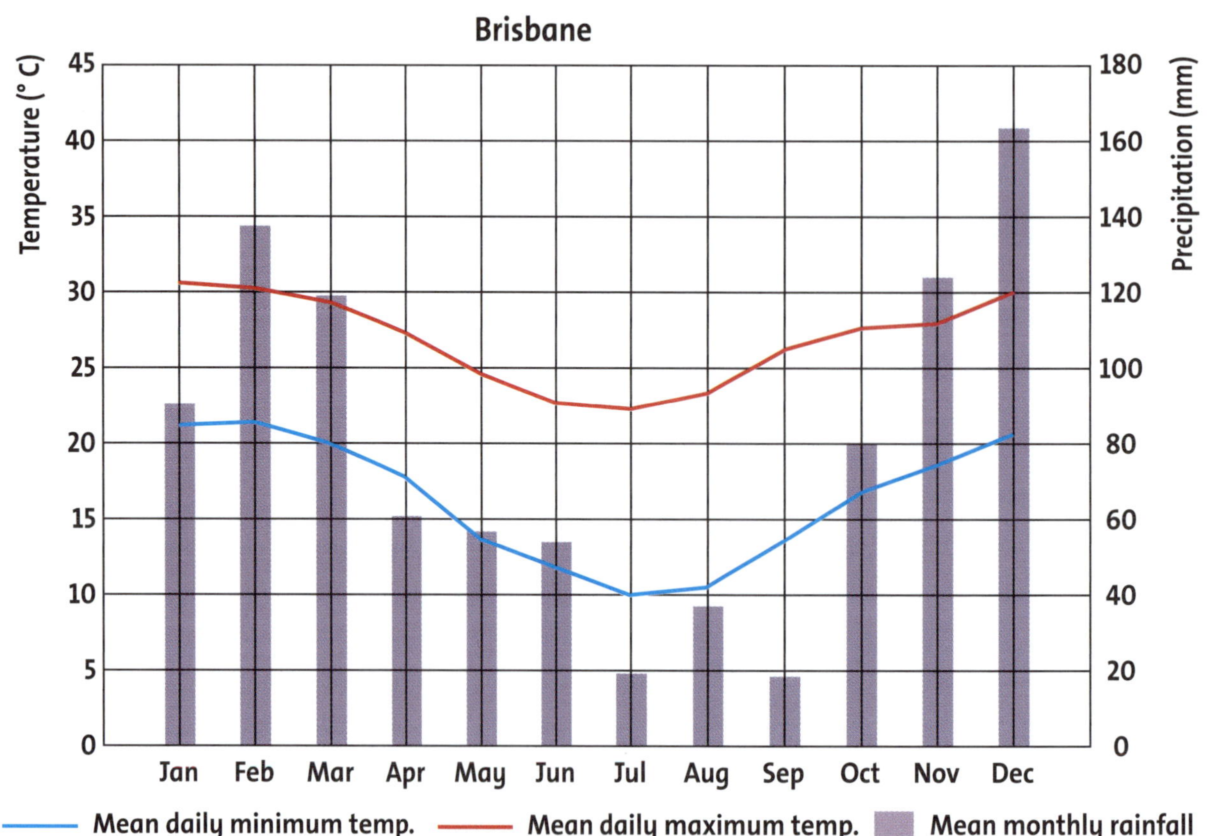

Source: http://singaporefun.weebly.com/uploads/3/0/0/2/30028935/8342379.jpg?656
http://www.farmonlineweather.com.au/images/climate/wz_clim_annual_site_40913.png]

GEOGRAPHY

Research

Singapore is one of Australia's neighbours. What is it like? Use an atlas, Google Earth or an encyclopedia to find out these facts.

Location: ______________________________

Population: ______________________________

Climate: ______________________________

Languages spoken: ______________________________

Currency: ______________________________

Key natural features:

____

____

Key human features:

____

____

Analysing

Use the graphs to complete the table.

Questions	Singapore	Brisbane
1 Where is it located in relation to where you live?		
2 Where is it in relation to the Equator?		
3 Where is it in relation to the Tropic of Capricorn?		
4 When is the coldest month?		
5 When is the hottest month?		
6 When is the wettest month? How much rain do they get?		
7 When is the driest month? How much rain do they get?		
8 When would be the best time of year to visit? Why?		

Questioning

Imagine you could talk with someone who lives in Singapore. What would you ask them about where they live? Write at least three questions.

____

____

____

____

____

____

____

Communicating

Record the temperature and rainfall at your school for a week. Measure the temperature in the morning and in the afternoon, at the same time each day. What do the results tell you?

Evaluating and Reflecting

Would you prefer to live in Brisbane or Singapore? Explain your reasons.

____

____

____

____

____

____

____

____

Places Profile: Rural Life

Australia is very urbanised – only about 10 per cent of people live outside major cities. Although Asia is becoming more urbanised, half of all Indonesians live in small towns and villages.

Australia

In Australia, reliable telecommunications and transport infrastructure mean that even people living in remote areas are connected to services. Trucks and aircraft carry goods to all parts of the country. Journeys to towns and cities may be long, but are still possible. Medical care is available via aerial services.

The challenge in many rural areas is long-term work opportunities. Large mining projects employ many people in remote areas of Australia, but often workers fly in to work and then return to cities.

Some Australian farmers drive computer-controlled tractors.

The Royal Flying Doctor Service takes healthcare to remote and rural Australians.

Did you know?
More than 65 million Indonesians don't have access to electricity.

Indonesia

Life in an Indonesian village is very different to Australian rural life. Roads and other transport infrastructure are basic or non-existent. People rarely, if ever, leave their villages.

Subsistence farming is the main work, largely done by hand and with livestock. Everyone in the family works, from a very young age. Along the Indonesian coast, fishing is the main industry.

Poverty in Indonesia has fallen for three decades but remains a challenge. In the remote eastern islands of Indonesia, 95 per cent of rural people are poor. They lack safe water and sanitation. Limited access to clean water means that villagers – particularly women and children – spend too much time fetching water.

Indonesian farmers drive 'cow-powered' tractors.

Fishermen repairing nets in Lamalera, Indonesia.

Source: Go Facts Geography: *Asia*, pp. 22–23, Blake Education

Research

The Royal Flying Doctor Service provides a very important service to people in remote and rural Australia. What are some key moments in its history? List at least four.

Questioning

What is the meaning of these words?

- Subsistence:
- Sanitation:
- Rural:
- Poverty:
- Remote:

Analysing

What role does technology have in the lives of people in rural Australia and rural Indonesia? How does this impact on their lives, and the way they interact with places?

Communicating

Prepare and present a short, 1–2 minute speech to your class, explaining the importance of **one** of these:

- mobile telecommunications to rural and remote communities in Australia and overseas
- aerial medical services
- safe drinking water.

Evaluating and Reflecting

Millions of people in Indonesia do not have access to electricity. How would your life be different without electricity?

UNIT 27

Making Decisions

Different countries have different ways of making important decisions for their citizens.

- In an **autocratic** country such as North Korea, one person has all the power to make decisions.
- In a **monarchy** such as Saudi Arabia, the decisions are made by the King.
- In a **communist** country such as China, the decisions are made by the members of a single political party.
- In a **democratic** country such as Australia, the people make the decisions through their elected representatives in parliament.

The word democracy comes from two Greek words: **demos** which means 'the people or citizens' and **kratos** which means 'power or rule'. So, **democracy** means 'rule by the people' and it has existed since ancient Greece. Each year, five hundred citizens who were free men were selected to serve for twelve months.

Women, children and slaves were not allowed. Once a decision was made, all the citizens had to vote by going to the assembly place. If the majority of citizens agreed with the decision, then it became a law. This is known as **direct democracy**.

In Australia we use a system known as **representative democracy**. Our adult citizens vote for a person to represent them in parliament. These representatives then make decisions to create and change the laws for everyone. This usually happens every three to four years at a state level and at a national (federal) level.

Illustrator: Paul Lennon

TARGETING HASS 3 © PASCAL PRESS ISBN: 9781925726046

Research

Find out about your State parliament.

How many representatives are elected?

____

How often are elections held?

____

What is the leader called?

____

What are the State houses of parliament called?

____

____

Questioning

If you could vote in a State election, what questions would you ask the candidates to find out who is the best one to represent you? Write five questions.

1 ______________________________

____

2 ______________________________

____

3 ______________________________

____

4 ______________________________

____

5 ______________________________

____

Analysing

In the ancient Greek method of direct democracy, every free man voted on decisions.

Do you think this method would work in Australia today? Explain your answer.

____

____

____

____

____

____

____

Communicating

Imagine you want to represent your class and make decisions about your school.

What would you say to your fellow students to encourage them to vote for you in an election?

Prepare a short, one-minute candidate speech and share it with a friend. Were they persuaded? Why or why not?

Evaluating and Reflecting

Which decision-making method do you think works best – autocracy or democracy? Explain your answer by giving some examples from your own experiences.

____

____

____

____

UNIT 28

People Power

Have you ever thought that a rule or decision was unfair? Or that a powerful person was doing something that you didn't agree with? Well, there are a couple of different ways that you can show that you are unhappy.

One way is by having a peaceful protest. This action usually involves people marching in the street or meeting in an open space to talk about why they are unhappy and what they want done about it. Protestors often call out slogans using loud speakers and wave placards or banners with signs and messages on them.

Another way of showing that you are unhappy, or you want to see something changed, is to create a petition.

This can be on paper, where people add their names and signatures, or it can be online through websites such as **change.org**. To start a petition, you need to know what the issue is, who will support it, and what you want to happen. If enough people sign the petition, then you can make change happen.

Creating change can start with just one person. **Greta Thunberg** wanted her government to take more action to prevent climate change, so she stood outside the Swedish parliament holding up signs that called for school students to strike. Soon, many other young people joined in and the school climate strike movement spread around the world. More than one million students were involved in coordinated multi-city protests – people power in action! As a result, in 2019 Greta was nominated for the Nobel Peace Prize and named *Time* magazine's Person of the Year.

Photo: © Sebastian Czapnik, ID 20443814, Dreamstime.com

CIVICS AND CITIZENSHIP

TARGETING HASS 3 © PASCAL PRESS ISBN: 9781925726046

Research

Peaceful protests can be very powerful tools for change. In 2018, an 11-year-old girl named Molly Steer started a protest movement called 'Straw no More'. What impact did her actions have?

Questioning

Why do people choose to protest?

What is a **slogan**?

What is a **placard**?

What is a **signature**?

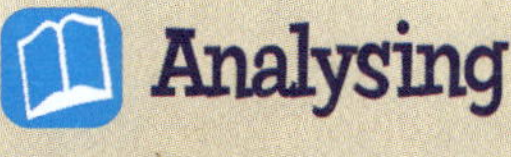

Analysing

Compare and contrast petitions and peaceful protests. How are they similar and different?

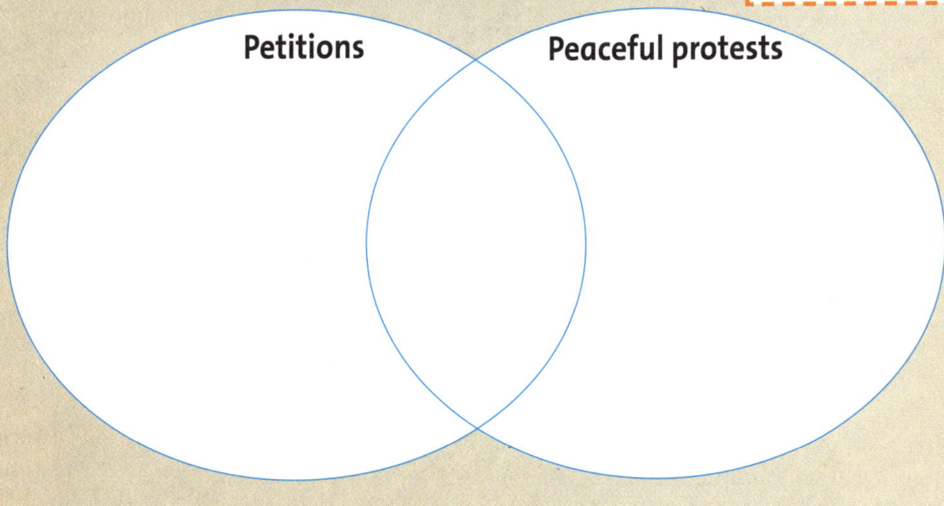

Communicating

Choose an issue that you feel strongly about.

Create a petition that could make a difference by asking people to take action.

Evaluating and Reflecting

Should students be allowed to go on strike and protest peacefully about climate change by marching in the streets? Why do you think this? Give at least two reasons.

UNIT 29

Story Places

The community of Jumbun is in North-East Queensland, and Jirbal, Girramay, Jiru and Gulngay peoples live there.

Their ancestors have lived in the area for thousands of years. They took only what they needed, living in balance with nature because the forest was their family, provider and teacher. Every generation followed the traditional law that told them who to marry, when to hunt and gather certain foods and how to settle conflicts. Law kept people healthy and at peace with each other, with the land and the ancestors watching over them.

The Aboriginal people remain closely connected with nature. They live with deep respect for the land and the spirits who have lived there since the beginning, before time. The Jumbun community thinks carefully about the wellbeing of the forest as they make decisions in their day-to-day lives. It is important to them that their children learn the old ways. They learn the skills and wisdom of tradition from their family.

Guyurru, a story place

At the top of Guyurru, the Murray Falls, are rocks carved out by the flowing water and clear pools shaded by leafy green trees. Guyurru is a story place.

The old people say that, a long time ago in the Dreamtime, creation spirits came and dreamt the forest and its creatures into being.

When the people came, they were given stories so that they would know how the mountains, animals and rivers came to be. The stories taught important lessons about living and how to dream life into nature every day by caring for the land.

The children of Jumbun learn these stories so that they can pass them on to future generations. Every place has a story to tell and something to teach.

Source: Sharing Country: *Rainforest*, p. 4, p. 16, p. 17, p. 24, p. 25, Steve Parish

CIVICS AND CITIZENSHIP

TARGETING HASS 3 © PASCAL PRESS ISBN: 9781925726046

Research

There are many Aboriginal Dreaming stories that describe what happens to people when they break the rules or act in a way that they shouldn't.

Research one of these stories. What is it called? What is its message?

Questioning

What makes Aboriginal story places special?

Do you have a special place where you like to listen to stories or read them yourself? Where is it and why is it special to you?

Why do Aboriginal elders tell Dreaming stories to very young children?

Analysing

"We have too many rules at school."
Do you think that this statement is true or false? Explain your answer.

Communicating

Think of three rules that you believe are important for everyone to follow. Discuss this in a small group.

What are the rules? Why are they important? What should the consequences be when the rules are broken?

Evaluating and Reflecting

Should people be punished every time they break a rule? Why or why not?

The Lion and the Mouse

Huge vines wrapped themselves around the bent and twisted branches of an ancient baobab tree. The leaves were so close together that they blocked out most of the sunlight. Only a few thin rays managed to stretch their way to the forest floor where the king of the jungle lay snoozing.

A small mouse crept towards the lion. "Oooh, what a lovely mane!" thought the mouse. "I wonder what it feels like. If I'm quick, he'll never even know that I'm here." And with that, the mouse scampered up onto the lion's tail, ran across his back and dived into the big fluffy mane.

But just then the lion yawned, and the mouse tumbled to the ground. The lion, now awake, grabbed the mouse and lifted him up by his tail.

"P ... pl ... please Mr Lion," said the mouse. "Let me go and one day I will come back and help you."

The lion roared with laughter. "You? Help me? Why, I am the King of the Jungle and you are so small and tiny. How could you ever help me?"

Yet the lion saw how frightened the little mouse was. And since he wasn't really hungry, he released his grip and let the mouse go.

Many weeks later, the lion was stalking some prey when he became caught in a net. The more he struggled to get free, the more stuck he became, until he couldn't even twitch his tail.

The little mouse heard the commotion. When he saw the lion trapped in the net, he went to work. Nibble, nibble, nibble ... Nibble, nibble, nibble. Bit by bit the little mouse gnawed away at the net until finally the lion was free.

"Dear mouse," said the lion. "I was wrong to make fun of your size. You have helped me after all – you've saved my life!"

Never forget: even the smallest friend is special.

Illustrator: Paul Lennon

TARGETING HASS 3 © PASCAL PRESS ISBN: 9781925726046

Research

What is a fable?

Find out the titles of three other fables and their messages to the reader.

1

2

3

Questioning

Why did the mouse scamper across the lion's back?

Why did the lion grab the mouse by his tail?

Why did the lion roar with laughter?

How did the lion know that the mouse was frightened?

What is the moral of the story?

Analysing

In the story, *The Lion and the Mouse*, the moral encourages people to be helpful and kind to others.

Is this a good rule? Why? Are there times when this rule would not work? What are they?

Communicating

Think of an important rule that young children should obey. For example: hold the hand of an adult when crossing the road.

Create a short imaginative story to share that rule.

Evaluating and Reflecting

Did the lion and the mouse treat each other fairly? Should all rules be fair? What does fairness look like and sound like?

Bush Regeneration: My Day

Andrew is a bush regeneration worker.

I think about the day ahead as I ride my bike to work, watching the sun rise. My team is going to work on some land overgrown with weeds. It's our job to remove them, so the native plants can grow again.

It's hard work so smiles are welcome!

I put on gloves and we spend the morning pulling out weeds. We leave them in piles to decompose. Lizards and insects will make new homes in the piles.
As the day warms up, we stop for a cup of tea. Then we move into the shade to chop back privet, which is a large weed. We put poison on the tips to stop it regrowing.

We pile up the sticks of dead weeds to make habitats for animals.

A couple of the guys plant native grasses along a creek.

After lunch, I plant some native trees. As I mulch and water them, I spot a red-bellied black snake. It is probably looking for the tree frogs I heard earlier. The rest of the team sprinkles native seed around. Late in the afternoon, we pack up our tools, jump on our bikes and head home.

A little whip snake we found. We'll leave him alone.

Source: Go Facts Geography: *Land, Air and Water*, pp. 12–13, Blake Education

CIVICS AND CITIZENSHIP

TARGETING HASS 3 © PASCAL PRESS ISBN: 9781925726046

Research

Andrew participates in his community by helping to regenerate native bush. What special community groups can people join in your local area?

Questioning

What is the meaning of these words?

- regeneration
- decompose
- mulch
- native

Analysing

What are the benefits of Andrew's actions?

Communicating

Do you have an idea for how you could participate in your community to help others or the environment?

Discuss it with others to get their feedback.

Evaluating and Reflecting

Should it be compulsory (the rule, required by law) for everyone to help others in their community? Explain why.

Caring for Places

People care for places that are important to them.

Kakadu has about 200 000 visitors each year.

Did you know?
More than 10 000 crocodiles live in Kakadu National Park.

People protect places so that future generations can learn about them and enjoy them. The National Trust in Australia began when people wanted to stop old buildings in Sydney from being knocked down. Now the National Trust looks after places such as Old Government House in Parramatta, the oldest public building in the country.

Volunteers work to protect penguins in Manly, New South Wales.

Picking up litter is caring for a place.

National parks are places set aside to preserve nature. Parks Australia protects some of Australia's most important natural places. Kakadu is the largest national park in Australia. Aboriginal people have looked after this place for more than 50 000 years. Now, they work with Parks Australia to manage the land.

Coastcare groups around Australia look after sections of coastline. Volunteers remove weeds and plant native plants. They teach people about native plants and animals.

Source: Go Facts Geography: *People and Places*, pp. 14–15, Blake Education

Research

What national parks are in your state or territory? How do people care for them?

Questioning

Imagine that you are interviewing one of the Coastcare volunteers. Write four questions that you would ask.

1

2

3

4

Analysing

What are the positive impacts of people caring for places that are important to them?

Communicating

Design a sign for a local beach or park to encourage visitors to clean up their rubbish instead of littering.

Evaluating and Reflecting

Think about your own school community. What is something that you and others can do to look after it and care for it? Why is this important?

History Assessment 1

What is the nature of the contribution made by different groups and individuals in the community?

1 Create a coat of arms to represent a particular person or group. For example: you, your family, your class, school, sporting team or dance troupe. Write a word or phrase on the banner to represent your person or group.

2 Explain your choice of words and images. How do they represent the contribution made to the wider community by the group or individual?

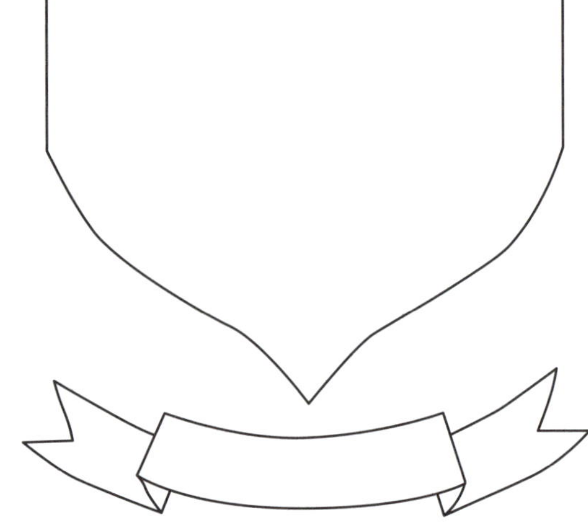

3 What is Sorry Day, and why is it important?

4 What is Harmony Day, and why is it important?

5 What do Sorry Day and Harmony Day have in common?

6 What symbols and emblems are used by these days, and what do they represent?

Sorry Day Symbols and Emblems	Harmony Day Symbols and Emblems

ASSESSMENT

TARGETING HASS 3 © PASCAL PRESS ISBN: 9781925726046

History Assessment 2

How and why do people choose to remember significant events of the past?

26 January 1788. Captain Arthur Phillip raises the British flag for the new convict settlement at Port Jackson in Sydney.
Illustration by Paul Lennon, based on a painting by Algernon Talmadge.

1 What does Australia Day mean to you?

____

____

____

2 Who might have a different opinion about Australia Day and what would it be?

____

____

____

3 Do you think that Australia Day should be celebrated or commemorated?
Why do you think this?

____

____

____

History Assessment 3

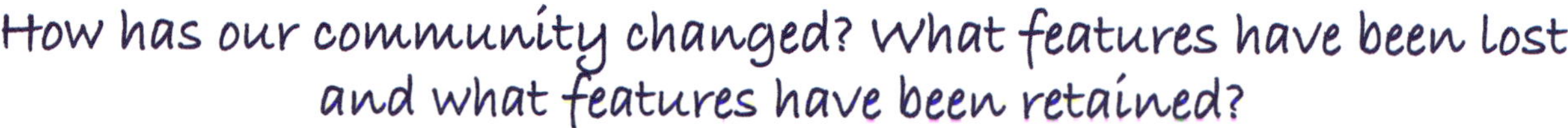

Task

Look carefully at the images. They are both of the same place – Sydney Cove.
Explain how this place has changed over time, and discuss the impact of those changes.

Sydney Cove 1803
Sydney from the west side of the cove. Artist: George Evans (1802)

Sydney Cove 2019

Sydney Cove – Then	Sydney Cove – Now

1 Why has Sydney Cove changed in these ways?

2 Overall, have these changes been mostly positive or mostly negative? Explain your answer.

ASSESSMENT

TARGETING HASS 3 © PASCAL PRESS ISBN: 9781925726046

Geography Assessment 1

What are the main natural and human features of Australia?

Tasks

On the map of Australia, show:

- the state and territory borders
- the state and territory capitals
- the Australian capital
- oceans and seas
- at least five natural features
- at least five human features.

Key

Geography Assessment 2

What would it be like to live in a neighbouring country?

South-east Asia and Australia

Tasks

1 On the map, show:

- the names of countries in the Asia-Pacific region
- the names of the oceans and seas
- the equator
- the Tropic of Cancer
- the Tropic of Capricorn
- at least five significant natural features
- the cardinal directions

2 If you could choose to live in any country in the Asia-Pacific region, where would you live? Why?

____

____

____

TARGETING HASS 3 © PASCAL PRESS ISBN: 9781925726046

Geography Assessment 3

How and why are places similar and different?

Tasks

1 Use the grid to design a special place. It could be a new classroom or school, or an outside recreational area.

Include a legend to explain the symbols and images you use.

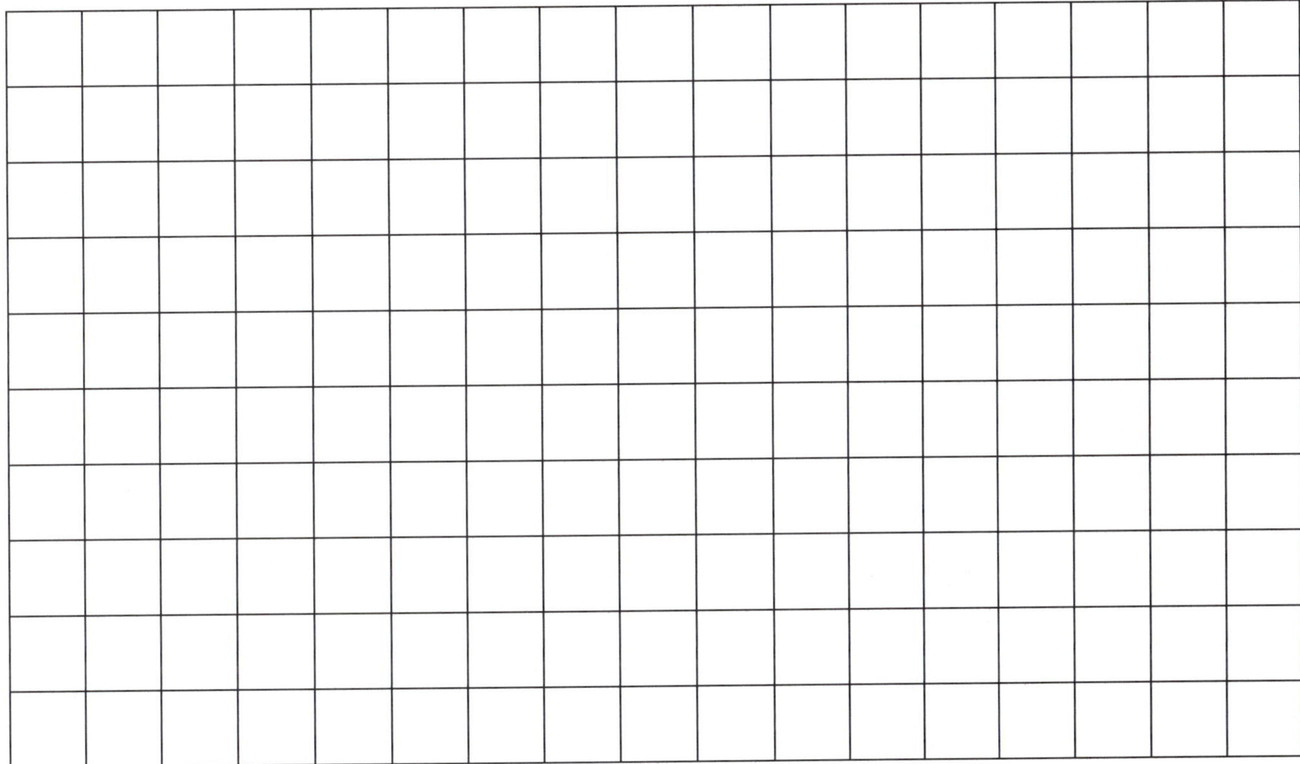

Legend

2 What special place have you designed? ____________________

3 Why did you choose this place?

____

4 Explain two of its special features.

____

____

Civics and citizenship Assessment 1

How do we make rules?

Tasks

Look at the images in the chart. For each image, explain what the rule is, why it is important and what the consequences are for breaking the rule.

Sign	What is the rule?	Why is this rule important?	Consequences if the rule is broken

ASSESSMENT

TARGETING HASS 3 © PASCAL PRESS ISBN: 9781925726046

Civics and citizenship Assessment 2

How do I participate in my community?

Tasks

1 Explain what the members of each type of group have in common.

2 Draw lines to match the types of community groups with examples. Some examples may belong to more than one group.

Members of a cultural group

____

____

____

Members of a sporting group

____

____

____

Members of a social group

____

____

____

Cultural

Sporting

Social

- Karate
- Dance
- Choir
- Lego club
- Book club
- Drama
- Football
- Chess
- Gardening group

3 Draw yourself participating in a community activity. What are you doing? Why are you doing it?

____

____

____

____

____

____

____

____

Answers

Note: Answers are not supplied to open-ended questions, where students' responses will vary.

ANSWERS

1 Home

Analysing: 1. false, 2. false, 3. true, 4. false, 5. true

Evaluating and reflecting: The extract gives the text authority and makes it more believable because it is first-hand evidence (a primary source).

2 First Peoples: Queensland and South Australia

Questioning:

What do the Aboriginal peoples of Queensland refer to themselves as?

What is another name for the Guugu Yimithirr people?

What is fire stick farming an example of?

Analysing:

Then (30 000 years ago)	Now
Cooler temperatures	Warmer
Covered in forests and grasslands	Much of the land has been cleared
Home of the Murri people	Home to many different groups of people

Evaluating and reflecting: Yes, with reasons related to its sustainable use, its ability to aid hunting and modify or change the landscape to create new grazing land for animals which meant that it helped the supply of food. Planning and organisation answers would involve wind direction and strength, making sure there were suitable fire breaks in place, that they had enough people to control it so it did not get out of control and that they did it in a suitable area.

3 Dreaming Story: Why the crocodile rolls

Analysing: Answers will vary, but may include responses such as punishing Min-na-wee, teaching her a lesson, showing her how her actions made other people feel, encouraging her to be nicer.

Evaluating and reflecting: Yes, because she would not have wanted to end up like that. Answers for the second part of the question will vary.

4 Looking after country

Research: The role of the elders can vary between different communities. They teach respect for the natural world and the spiritual relationship between the Aboriginal people and the land, feeling the rhythms of the elements and the seasons. They care for grandchildren, sit in local Koori or Murri courts and discuss important community issues.

Analysing: The sharing of art and stories are important ways of sharing a person's feelings and experiences. They help us to understand the world around us. If we did not have stories and art, then we would find it more difficult to communicate with other people and share ideas, especially between people from different cultures, backgrounds and religions.

5 Sydney Harbour Bridge

Analysing: Sample responses may include:

1930: being built, incomplete, pedestrians, carriages, celebrated, not many tall buildings around it

Now: complete, busy, lots of traffic, cars and trucks, trains, lots of tall buildings nearby

1930 and Now: cars travel across, sandstone pylons, rail line, multiple lanes north and south

Evaluating and reflecting: Answers will vary, but sample responses may include:

- an unusual shape or design
- it involved a lot of people to create or build
- it cost a lot of money
- lots of people use it regularly
- it is in a prominent position
- it is used in advertising for tourists.

We give structures nicknames because they are funny and because we like them.

6 The Snowy Mountain Scheme

Research: Answers will vary, but some sample responses may include:

- being far from home without friends and family
- not everyone spoke the same language
- they did not understand the customs and beliefs of people in their new country
- some suffered post-traumatic stress from their experiences in Europe
- they didn't have much money
- they lived in temporary houses.

Analysing: The text tells us that life was difficult. Many people came to Australia from all over Europe and they were trying to start new lives. The working conditions were harsh. Lots of people lived in temporary camps. Australia was trying to develop itself and grow by building projects such as the Snowy Mountains Scheme.

Evaluating and reflecting: Answers will vary, but responses may include:

- they had settled in here
- they didn't have homes and families in Europe to return to
- they were happy here
- they didn't want to leave their new friends and jobs.

7 An Interview with my Grandad

Evaluating and reflecting: Answers will vary, but may include

- things now cost more at the canteen
- more people drive to school in a car
- there are now fewer children in each class
- we don't learn by chanting now
- we don't get given bottles of milk at school now
- there are more healthy options at the canteen
- children do not get caned as a punishment, there is detention instead.

8 Victor Chang

Research: Heart disease is damage caused to the major blood vessels where they become narrow and blocked. It leads to pain, heart attack and angina. It is caused by high blood pressure, smoking, excess alcohol consumption, diabetes, poor diet and lack of exercise. It can be prevented by regular exercise and physical activity, controlling your weight, lowering your cholesterol and controlling your risk of diabetes.

Questioning: Answers will vary, but some possible responses are:

- When did Victor Chang start the transplant program?
- What concerned Victor Chang about the program?
- Why was he voted Australian of the Century?
- Why did he become a doctor?
- Who helped Victor Chang develop an artificial heart?
- How is he remembered today?

Answers

Note: Answers are not supplied to open-ended questions, where students' responses will vary.

Analysing: Possible responses include:
- He saved many lives and demonstrated commitment and determination to achieve his goal.
- He made a significant contribution to Australian society as well as to medical research.

Evaluating and reflecting: Answers will vary, but responses may include:
- helping others not just themselves
- being a good role model
- people who strive for excellence in their field/job/sport/passion
- they inspire others with their achievements.

9 Ash Barty

Research: Evonne Goolagong Cawley was an Australian former World Number 1 tennis player. She won 14 grand slam singles titles. She is a member of the Wiradjuri people of NSW. She retired in 1983. She was inducted into the International Tennis Hall of Fame in 1988. She has also been a member of the Australian Sports Commission and a Sports Ambassador to ATSI communities. She was Australian of the Year in 1971 and received an MBE (Member of the Order of the British Empire) in 1972.

Questioning: 1. false, 2. true, 3. true, 4. true, 5. true, 6. false

Analysing: Ash has promoted Indigenous sporting participation. She is the National Indigenous Tennis Ambassador. She wants to inspire others and encourage young people to get active. This is important because there are not a lot of modern Indigenous role models, especially females.

10 Symbols and celebrations

Research: The Union Jack represents the history of British settlement. The Federation star has seven points which represent the six states and the territories of Australia. The Southern Cross is a constellation of five stars that can only be seen in the southern hemisphere and is a symbol of our geography. It was also used as the flag during the Eureka stockade rebellion.

Questioning:

The coat of arms represents the whole country through the native kangaroo and emu, as well as our national flower, the wattle, and the Federation star on the top.

The coat of arms represents the fact that the states are joined together in one nation. Each state is represented by a section of the shield.

The coat of arms does not really represent our long history of Aboriginal culture unless you include the native flora and fauna which were part of some Indigenous groups' diets.

Analysing: National symbols are important because they represent the respect we have for our country and the people in it. They are a way to identify us around the world and show that we are unique and special. They also show others what values we have as a nation.

Communicating:

1878 – *Advance Australia Fair* was written.

1899 – The Australian cricket team wore green and gold caps.

1901 – Australia became a nation and the Australian flag was first flown.

1908 – The first coat of arms was introduced.

1912 – The second coat of arms (the one we use now) was introduced.

1953 – The Australian flag was proclaimed the official national flag.

1984 – *Advance Australia Fair* became Australia's official national anthem.

1984 – Green and gold became our official national colours.

11 Sorry Day and Harmony Day

Analysing: The mistakes were:

1. The British took land belonging to Aboriginal and Torres Strait Islander peoples.
2. The British killed or hurt Aboriginal and Torres Strait Islander peoples.
3. The government took many Aboriginal and Torres Strait Islander children from their families.

The answers for the second part of the question will vary, but may include how important it is to apologise and say sorry when you have done something wrong.

12 Vietnamese New year (Tet)

Research:

Similarities	Differences
Usually celebrated on the same day Based on the lunar calendar People celebrate with food, music and dance Houses are cleaned Spending time with grandparents and family is important Dragon dances	Vietnamese call it *Tet.* Before the day, Chinese families paste red scrolls around their houses with words such as *peaceful* and *safe* to give them good luck. Tet lasts from three days to a week. Chinese New Year lasts from three days to two weeks. Vietnamese people eat *mut* (candied fruits) and *banh chung* (cake of sticky rice and beans) . Vietnamese people say *Chuc Mung Nam Moi* (happy new year). Chinese people say *Gong hei fat choy* in Cantonese, or *gong xi fa cai* in Mandarin. Each new year is known for an animal of the Chinese zodiac.

13 Ramadan

Questioning:

Ramadan begins when a crescent moon appears in the sky. (It is also the 9th month of the Islamic calendar.)

The meal Muslims eat before sunrise is *suhoor.*

Muslims show their thanks by praying and sharing their money with the poor.

Eid-ul-fitr means Festival of breaking the fast.

The Muslim place of worship is called a mosque.

The Koran is the holy book of Islam.

Analysing: Responses may include a reference to children, pregnant and sick people not being physically able to avoid eating. It would be unhealthy for them to fast and it might make them sick.

14 Anzac Day

Research: The red poppy started as a Remembrance Day symbol but has spread to be included in Anzac Day services and ceremonies. It represents the flowers which grew over the battlefields of France in WWI. People wear them to remember those who died in war or who still serve.

Evaluating and reflecting: Answers will vary, but responses may include references to how important it is to remember those who fought for our country and who made the ultimate sacrifice. It was an important moment for a young country such as Australia and it helped us define ourselves as a nation. It is important to learn from things that happened in the past so perhaps we will not have wars like that in the future.

Answers

Note: Answers are not supplied to open-ended questions, where students' responses will vary.

15 Australian Landscapes

Research: Lake Eyre is Australia's largest salt lake but it usually contains little or no water. It is found in the arid and semi-arid deserts of Central Australia. Its Aboriginal name is Kati Thanda. When it rains in the north, the waters flow down into the lake. This happens only a few times each century. The lake is Australia's lowest point below sea level. The lake's basin holds some of the rarest ecosystems in the world.

Questioning:

A temperate environment has four different seasons. It is moderate and does not have extreme ranges in temperatures.

An arid environment is very dry and has low levels of rainfall. It is also very hot.

A reservoir is a large body of water (natural or constructed) used as a water supply.

Mt Kosciuszko is 2228 m high.

The Great Dividing Range crosses through Victoria, New South Wales and Queensland.

The Great Barrier Reef is special because it is the world's largest coral reef system.

Analysing: Uluru is sometimes known as the heart of Australia because it is almost in the centre of the country, like your heart is almost in the centre of your chest.

16 States and Territories

Research:

NSW: 1  2 waratah, 3 kookaburra, 4 platypus, 5 eastern blue groper, 6 black opal, 7 sky blue, 8 Newly risen how brightly you shine

Vic.: 1  2 common (pink) heath, 3 Leadbeater's possum, 4 helmeted honeyeater, 5 weedy sea dragon, 6 gold, 7 navy blue and white, 8 Peace and prosperity

Qld: 1  2 Cooktown orchid, 3 koala, 4 brolga, 5 barrier reef anemone fish, 6 sapphire, 7 maroon, 8 Bold but faithful

SA: 1 2 Sturt's desert pea, 3 hairy nosed wombat, 4 piping shrike, 5 leafy sea dragon, 6 opal, 7 blue, red and gold, 8 The festival state; The wine state

WA: 1  2 red and green kangaroo paw, 3 numbat, 4 black swan, 5 whale shark, 6 (none), 7 black and gold, 8 The wildflower state; The golden state

Tas.: 1  2 Tasmanian blue gum, 3 Tasmanian devil, 4 yellow wattlebird, 5 (none), 6 crocoite, 7 bottle green, yellow and maroon, 8 Fertility and faithfulness

ACT: 1  2 royal bluebell, 3 southern brush-tailed rock wallaby, 4 gang gang cockatoo, 5 (none), 6 (none), 7 blue and gold, 8 For the Queen, the law and the people

NT: 1  2 Sturt's desert rose, 3 red kangaroo, 4 wedge-tailed eagle, 5 (none), 6 (none), 7 black, white and red ochre, 8 The top end

Questioning: Answers will vary, but responses may include references to it being a long way from Holland, that the landscape was harsh and access was difficult.

Analysing: Each capital city is near the coastline. This is because there is a supply of fresh water. It is also because early exploration occurred by ship and these places are good ports.

Communicating:

1788 – British colonisation (New South Wales)

1825 – The southern island was made the colony of Van Diemen's Land.

1829 – Swan River Colony established (Western Australia).

1832 – Colony of Western Australia proclaimed.

1836 – Province of South Australia

1851 – Colony of Victoria

1856 – Van Diemen's Land was renamed Tasmania.

1859 – Colony of Queensland

1863 – The Northern Territory became part of the colony of South Australia.

1901 – Federation of Australia as a nation

1911 – Federal Capital Territory

1911 – The Northern Territory separated from South Australia.

1938 – The Federal Capital Territory became known as the Australian Capital Territory.

17 Lost cities of stone

Research: Answers will vary, but responses may include:

- They were formed by limestone.
- They are Western Australia's most-visited attraction.
- Some of the pinnacles are up to 5 m high.
- They were formed approximately 25 000 to 30 000 years ago.
- They were made from sea shells.
- They are in the Nambung National Park.
- They are different shapes and sizes.
- Aboriginal artefacts in the area date back more than 6000 years.
- The area is home to the Yuart tribe.

Analysing: Responses will vary, but may include references to the fact that it is surrounded by thick forests and is hard to get to. The lost city is so tall that up close you wouldn't see much of it. Some challenges would be walking up the steep mountains, getting up the sheer rock faces of the cliffs, getting through the thick forest surrounding it and getting to the area in the first place because it is not near major towns or cities.

Evaluating and reflecting: Answers will vary, but may include references to the fact that different groups having different views about the area would make it difficult to decide how the land should be used and who should be allowed to use it. Each group would want to follow their own ideas and have control. It would make protecting the area more important because of these competing interests.

18 Old Government House Parramatta

Research:

Governor	Dates served
Captain Arthur Phillip	1788–1792
Captain John Hunter	1795–1800
Captain Phillip King	1800–1806
Captain William Bligh	1806–1808
Major-General Lachlan Macquarie	1810–1821
Major-General Sir Thomas Brisbane	1821–1825
Lieutenant-General Sir Ralph Darling	1825–1831
Major-General Sir Richard Bourke	1831–1837
Sir George Gipps	1838–1846
Sir Charles Augustus Fitzroy	1846–1855

Questioning: Example questions could include:

- Who was the last governor to live in old government house?
- When was it built?
- Who built it?
- Why was it built?
- How has the building changed over the years?
- What special protection does old government house have?

Analysing: World Heritage sites are constructed or natural areas or structures that have outstanding cultural, historical, scientific or other significance and deserve special protection.

TARGETING HASS 3 © PASCAL PRESS ISBN: 9781925726046

Answers

Note: Answers are not supplied to open-ended questions, where students' responses will vary.

19 Australia's neighbours in the Asia-Pacific

Questioning: 1. Japan, 2. New Zealand, 3. Vietnam, 4. China, 5. Papua New Guinea

Analysing: Many of the countries are made up of multiple islands. This is important because it changes the way that people interact with each other. It means that fishing is an important food source. It also changes the way we trade with them. It is also important because with rising sea levels, these countries will face problems in the future that Australia may be able to help with.

Evaluating and reflecting: Answers will vary, but responses could include that ASEAN is important because there are lots of smaller nations in the Asia-Pacific region and their people are spread out over different islands. When these countries are so close and will be impacted in similar ways by changes in climate and weather, it is important that neighbours help each other.

20 More than one Asia

Research: Answers will vary, but may include:
- Angkor Wat is the largest religious monument in the world.
- It was originally a Hindu temple.
- It was dedicated to the god Vishnu.
- It became a Buddhist temple at the end of the 12th century
- King Suryavarman II of the Khmer Empire built Angkor Wat.
- It is in Cambodia.
- It is pictured on the national flag.
- Its name means 'city of temples'.

Questioning: 1. false, 2. true, 3. false, 4. false, 5. true, 6. false, 7. true, 8. true

21 Maori

Research: The haka is an ancient Māori war dance. It was traditionally used on the battlefield. It was also used when different groups came together peacefully. It is a display of a tribe's pride, unity and strength. When performing the haka, dancers will stamp their feet, stick out their tongue, slap their body in different rhythms and chant loudly. In modern times, the haka is used at special ceremonies and celebrations and to challenge teams on the sporting field.

Analysing: The carvings tell a story and record historical events. Some are mythical or spiritual beings such as Marakihau.

Evaluating and reflecting: Answers will vary, but may reflect upon the more war-like nature of the Māori people, or that the Aboriginal peoples were more diverse in terms of cultural groups and language. The Māori appeared to be more 'advanced' to the British settlers in terms of weapons and tools so they were better able to protect themselves.

22 My Asia

Research:

Capital city: Tokyo
Four main islands: Hokkaido, Honshu, Shikoku, Kyushu
Currency: Japanese Yen
Highest mountain: Mount Fuji
Native animals include: Amami rabbit, Japanese macaque, sika deer, green pheasant, Japanese giant salamander, leopard cat.

Questioning: Possible answers include:
- Nara: Where was Kenji born in Japan?
- Multiculturalism: What term refers to people from different cultures who live together in a community?
- Sushi: What is a popular Japanese dish made with rice and vegetables?
- Ceramicist: What is a person who makes pottery items called?

23 People, Places and Spaces

Questioning: Answers will vary, but may include responses such as:
- What are some natural features of places?
- What are some human features of places?
- Why do places have different meanings for different people?
- How is a public place different to a private place?
- What is an example of a public place?
- What is an example of a private place?
- Why do people name places?

25 Places are similar and different

Research:

Location: South-East Asia, just north of the equator, on the southern border of Malaysia

Population: 5.6 million

Climate: Warm and tropical all year round. It has two monsoonal seasons with heavy rainfall separated by a drier inter-monsoon period. The temperature is consistent all year round.

Languages spoken: English, Mandarin, Malay

Currency: Singapore Dollar

Key natural features: Lazarus beach, Bukit Timah nature reserve, Kusu island, Labrador park, Mount Faber, The southern ridges, MacRitchie nature trail

Key human features: Helix bridge, Raffles Hotel, Gardens by the Bay, Marina Bay Sands, Clarke Quay, Sentosa island

Analysing:

	Singapore	Brisbane
1	Answers will vary.	Answers will vary.
2	north	south
3	north	south
4	December/ January	July
5	April/May/June	January
6	December 150 mm	December 18 mm
7	260 mm	145 mm
8	July, because there is less rain and not maximum temperatures.	July–Sept because there are warm temperatures and less rain.

26 Places Profile: Rural Life

Research: Answers may include:
- The Royal Flying Doctor Service began as the Australian Inland Mission Aerial Medical Service.
- It started in Queensland in 1928.
- By 1929, people used pedal radios to contact them.
- The service began in Victoria in 1934.
- The Western Australian base was opened in 1935.
- The NSW branch formed in 1936.
- In 1955, Queen Elizabeth II visited the RFDS base in Broken Hill, NSW.
- Tasmania started their branch in 1960.
- In 1994, the Reverend John Flynn who began the RFDS was pictured on the $20 note.

Questioning:

Subsistence: to support yourself at a minimum level

Sanitation: conditions relating to public health such as clean drinking water and sewage disposal

Rural: relating to the countryside, not towns or cities

Poverty: not having enough money or possessions for your personal needs

Remote: far apart in distance, out of the way and secluded

Answers

Note: Answers are not supplied to open-ended questions, where students' responses will vary.

ANSWERS

Analysing: Answers will vary but may include aspects such as:
- In Australia, technology is used for communication and in the machinery used on farms. It is also used in transporting people and goods from one place to another.
- In Indonesia, the use of technology is much simpler. They use hand-held tools or animals to assist them in farming, not machines, and they rarely leave their farms. This means that more people are needed on farms in Indonesia because they do not use machines to help and harvesting the produce takes more time.

Evaluating and reflecting: Answers will vary, but responses may include references to not being able to communicate easily with friends and relatives, not being able to do homework, having to do more jobs manually, having to walk to school instead of driving or catching a bus, having to write things out instead of typing them, not being able to interact with others in computer games, having different hobbies, not watching movies on television and being outside more often because of this, and not always having hot water.

27 Making Decisions

Research: Answers will vary, but responses may include references to voting, having an election, numbering names on a ballot paper.

28 People Power

Research: Molly's campaign started at her school when she asked her principal if the canteen could stop using plastic straws. Her message quickly reached people all over the world. She saw the impact plastic drinking straws were having on animals and the environment and so she used the internet and news media to spread her concerns and solutions to others. Now Woolworths have decided to stop selling plastic straws and McDonalds in the United Kingdom have done the same. Many smaller businesses are also changing to paper straws.

Questioning:

People choose to protest when they disagree with something or they think a rule or decision is unfair.

A slogan is a short memorable phrase that is used to grab people's attention. It is repetitive and expresses an idea or goal.

A placard is a printed or handwritten sign that is displayed in public or carried during a protest march. It uses words and images to convey a message.

A signature is a person's name written in a distinctive way that can be recognised as theirs and not belonging to someone else.

Analysing:

Petitions:
- Only in written form
- Use names, signatures and addresses
- One person at a time
- Quiet
- Do not impact on people who are not involved

Petitions and peaceful protests:
- People asking for change
- Lots of people agreeing with an issue
- Not always successful

Peaceful protests:
- Verbal or physical (shouting and marching)
- Many people all at the same time
- Loud
- Can impact people who are not involved – traffic jams, people stopped from getting where they would like to go

29 Story Places

Questioning:

Aboriginal story places are special because of their connection to country and their significance to the people.

Aboriginal elders tell stories to the young children to teach them about Aboriginal history and to explain how the land came to be. They also tell stories to show children how to behave and that there are consequences for breaking the rules of the community.

30 The lion and the mouse

Research: A fable is a short story, often with animal characters, that shares a moral (an understanding of right and wrong behaviour).

Questioning:

The mouse scampered across the lion's back because he wanted to feel his mane.

The lion grabbed the mouse by the tail because the mouse had woken him up and he wanted to find out why.

The lion knew that the mouse was frightened because he was stammering.

The moral to the story is to be nice to other people because even the smallest friend is special.

Evaluating and reflecting: Yes, the lion and mouse treated each other fairly. Answers will vary.

31 Bush regeneration: My Day

Questioning:

Regeneration: bringing new life to an area, to help grow after loss or damage, to revive

Decompose: to decay or break down

Mulch: material such as leaves and bark spread around a plant to provide nutrients for the soil

Native: something indigenous to a particular place or area

Analysing: The benefits of Andrew's actions are that by removing the weeds, it encourages native plants to grow and this means that there are new homes for animals. It supports the whole ecosystem.

32 Caring for Places

Analysing: Answers will vary, but responses may include ideas about keeping places for future generations and enabling many people and groups to enjoy the place.

History Assessment 1

3. Sorry Day is a day to remember and commemorate the way Aboriginal and Torres Strait Islander peoples were treated. It is a day to promote reconciliation. It is important as it helps people understand what happened in the past and strive to make improvements in the future.
4. Harmony Day is a day to celebrate and promote diversity and the acceptance of people who may be different, so that everyone belongs. It is important because it helps to create a better understanding of other cultures and works to decrease discrimination.
5. Sorry Day and Harmony Day both focus on improving relationships between people of different cultures and backgrounds.

TARGETING HASS 3 © PASCAL PRESS ISBN: 9781925726046

Answers

Note: Answers are not supplied to open-ended questions, where students' responses will vary.

6.

Sorry Day	Harmony Day
• The Aboriginal Flag: the black symbolises the people, the red is the earth and the Aboriginal People's spiritual connection with it, and the yellow is the sun. • Torres Strait Islander flag: the green is the land, the blue is the sea, the black is the people, the white dharri are the Torres Strait People, and the star represents peace.	• The colour orange symbolises social communication and meaningful conversations as well as respecting each other. • Joined hands are a symbol of connectedness between different people.

History Assessment 2

2. Answers may include: Aboriginal and Torres Strait Islander peoples, new migrants, people who live in Australia but are not Australian citizens
3. Sample response for 'celebrated': It is a significant day of change in Australia's history and as such it should be celebrated. Australia is the way it is today because of these events on 26th January 1788. Similar days in other countries are celebrated, such as Bastille Day in France. It is a day where everyone can celebrate being part of a wonderful country.

   Sample response for 'commemorated': Australia Day should be commemorated because it is a day that led to great suffering for the Aboriginal and Torres Strait Islander peoples. They lost control of their land and their way of life. It is a day that should be remembered and not forgotten, but not celebrated with happiness and joy.

History Assessment 3

Answers may include:

- Then: short buildings, sailing boats on harbour, buildings have chimneys, buildings made of stone, buildings were long and rectangular
- Now: buildings tall (skyscrapers), buildings made of metal and glass, buildings have curves and triangular shapes, buildings are closer together, mix of old and modern

1. Improved technology and engineering so buildings can be taller, increased population, changes in the way the area is used, more accommodation for people living (apartments), visiting (hotels) and working (office towers) in the city

Geography Assessment 1

The human and natural features identified will vary.

Geography Assessment 2

The natural features identified will vary.

Civics and Citizenship Assessment 1

No dogs allowed.
- It keeps people safe and keeps the environment healthy.
- If you break the rule, there is a warning or a fine. People might get bitten by dogs or the dogs might poo on the ground.

No swimming.
- It keeps people safe.
- If the rule is broken, people might drown in dangerous conditions.

Do not feed the birds.
- It protects the environment and the health of the birds.
- Birds could get sick and die from eating the wrong food. It also encourages birds to eat/take human food which can make birds' actions annoying.

No littering.
- It protects the environment and can stop people from getting sick.
- Litter looks and smells bad. It can get into our waterways and cause pollution. Animals can eat the rubbish and get sick or die.

Do not enter/do not climb the fence.
- It keeps people safe.
- People could get hurt if they enter the area.

Civics and Citizenship Assessment 2

1. Members of a cultural group share a common cultural background. This may include a country or region where they or their family comes from, it may include the language they speak or their religious beliefs.

   Members of a sporting group share enjoyment by playing the same sport. They follow or support the same team or club.

   Members of a social group enjoy spending time with each other. They have similar interests or hobbies.
2. Cultural: dance, choir, drama

   Sporting: karate, football, chess, dance

   Social: dance, choir, Lego club, book club, drama, gardening group, chess

Targeting HASS
Year 3

Copyright © 2020 Pascal Press
ISBN: 978 1 925726 04 6
Reprinted 2022, 2025

Published by Pascal Press
PO Box 250
Glebe NSW 2037
contact@pascalpress.com.au

Author: Merryn Whitfield
Publisher: Lynn Dickinson
Typesetter: Ruth Schultz
Editor: Ruth Schultz
Designer: Janice Bowles

Printed by Vivar Printing/Green Giant Press